Orgonomic Sciences Handbook

*a brief narrated history of
Wilhelm Reich's greatest
discoveries*

An original manuscript of
dr. Leopoldo Orsini Corvetti

Distributed worldwide via Lulu
First Printing, December 2019
ISBN 978-0-244-54393-8
V1.0.1

More about the author on
holisticenergy.xyz

INTRODUCTION

Orgone energy plays a fundamental role in our lives yet only a tiny part of the population can perceive it due to natural physical differences. Most of us are unaware of constant streams of energy passing through our bodies with direct interferences from the biological level to the mental one. As shown by Wilhelm Reich himself, it is possible to interfere with this kind energy, to redirect it where needed and to purify it from the negative charge (or better said, to separate the negatively charged energy to the positive one) using appropriate filters and accumulators.

Schematically, Orgonomic sciences can be split into four main branches: physics, biology, medicine and sociology. If in the European period of Reich's life sociology (and related psychoanalysis) constituted the main element of his scientific research, in the American period the doctor focused on all the three disciplines: physics with the study of cosmic energies such as astral Orgone, biology with the study of the entities called bions found in living and organic tissues, medicine with the macroscopic effects of Orgone energy on organisms and the cancerous biopathy problem. All those main aspects of Orgonomic sciences will be briefly touched in the following pages.

This book is also one of the several memorials to this forgotten genius of the 20th century, persecuted for his disruptive ideas, dr. Wilhelm Reich. He was a man of substance, successful and bright, who was encaged by an underdeveloped culture rooted in old Christian values. He dared to promote sexual freedom as an actual medical therapy, collectively breaking a set of unwritten rules that used to repress erotic hedonism and pursue an unnatural sexual taboo. The main crime of Reich was to give speech to the human nature, which is made of both intelligent consciousness and instinctual emotions. Therefore, the term individual itself means indivisible. Leading the way out of a subculture of forced abstinence and outdated practices like circumcision and female chastity, Reich quickly became a dangerous threat for the contemporary American society. Orgonomic sciences were banned and Reich books burned, bringing back memories from the Holy Inquisition.

Nowadays, Orgone has been labelled as a "pseudoscience" and is banned from any academic institution. Luckily many independent researchers and laboratories are carrying out experiments and confirming Reich's hypothesis, showing how science needs to rediscover its origins of independent quest for knowledge, away from the influencing power of economy and politics.

But beware, there's lots of mysticism and infamous preaching on Orgone: because Orgonomic sciences have always been hidden from the public, many self-defined experts have been preaching ignoring Reich's works and giving the public a misleading and confused view full of technical mistakes, misconceptions and induced biases. Their aim is obvious, perpetrating the ban over Orgone in a new manner, i.e. by muddying the waters and comparing it to other unrelated theories such as the flat Earth or the Ancient Astronauts.

Nonetheless, as it has happened with traditional and holistic medicine too, a business opportunity always carries many avid profit-seeker individuals with no real interest in pursuing the well-being of people, who instead just want to flood the market with useless devices with untested and potentially dangerous effects. This specific topic was subject of an article I wrote on Medium® I strongly recommend taking a look at, *The Orgonite Scam*. It caused a strong wave of negative comments from worldwide untrusted Orgonomic devices sellers as their businesses are based on people's ignorance and anyone trying to unveil the mist of esotericism and fictional complexity around the topic is considered as a threat.

Taking into account such events and the several requests from many people who just wanted to be able to take advantage of Orgone energy to their own wellbeing, I decided to write this guide in order to allow anyone to build their own Orgone devices and test their efficiency at home without relying on doubtful sources. At the same time, if you want a professional to build your accumulator, you will be instructed on a few simple tests to carry on at home with limited equipment so that you can really say if you got what you paid for.

Nonetheless, any self-respecting guide must include a brief introduction to the topic, so does this book scraping out the real history of Orgonomics without the technicalities of a formal academic textbook yet written with the same spirit, to teach and to spread the knowledge about something that gathers us all: *Orgone*.

Remember, Orgone is not a joke, a misconfigured accumulator can cause severe damage to the body. The best way to benefit from the positive effects of the natural orgonic flux and prevent injuries is to build and test any device yourself. As for health problems, Orgone can help to maintain and recover a good health status but it is not a cure, in case of medical problems consult a physician.

THE HISTORY OF ORGONOMIC SCIENCES

A "NEW" TYPE OF ENERGY

Orgone energy is a form of energy pervading the whole Universe, highly interactive with organic matter. Its carrier particle is the *bion* that can have a positive or a negative charge, giving birth to either the *Positive Orgone* (POR) or the *Negative/Deadly Orgone* (DOR). Living beings can be affected by the presence of surrounding Orgone energy fields as well as actively change the bions carrier charge, from negative to positive or vice versa. Such effect is quite similar to polarisation in physics, where light vibrating on various planes can be filtered to vibrate only on a chosen axis. The same happens to Orgone, where the negative mindset of an individual can filter out all the POR and leave out just the negatively charged one. In fact, strong emotions can interfere with the Orgone energy flow and cause an abrupt transition from one polarisation state to the other, with a subsequent switch in the carrier charge of the bions. It has also been shown in several experiments that organic materials hold Orgone whereas metallic ones attract and repel it in a short time.

Orgone can also be highly affected by other physical forms of energy like electromagnetic and nuclear radiation, an effect that was demonstrated in the Oranur experiment.

New discoveries in physics also lead the path to the obvious necessity of including bions and Orgone in the pantheon of particles composing the Universe: dark matter and dark energy are just forms of WIMP (*weakly interactive mass particles*) found around galaxies, very difficult to detect using traditional experimental apparatuses especially due to the fact that scientists themselves don't know what to look for.

Yet the existence of a cosmic pervasive energy was well known since ancient times. In India, it was called *Prana,* in the Jewish Kabbalah it was called *Yesod* or *Astral Light,* an idea the Cristian culture later borrowed depicting Jesus and the Saints with a halo of glowing light. Meanwhile the Chinese, looking at the distribution of energy within organisms and the way a blockage may cause illness, developed the concept of life force, or *Chi.* Pythagoras around 500BC also spoke about a life energy as a luminous substance that could induce tangible effects on the human body. Paracelsus many centuries later called it *Arqueo o Munia,* defining it as a vital force and comparing it to an irradiating essence with beneficial effect on human health.

THE ETHER PROBLEM

The medieval concept of the universe was based on a mixture of five elements: earth, fire, air, water and ether. Concentric spheres divided the space into different regions. The innermost spheres are the terrestrial spheres bound to the human beings' influence, while the outer are celestial ones made of ether and contain the stars.

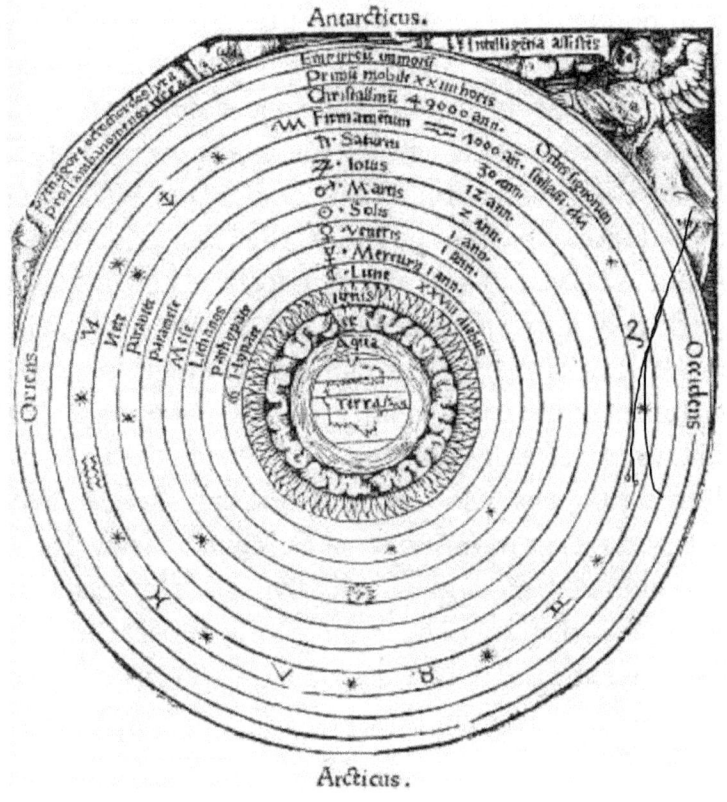

The word αἰθήρ (*aithēr*) in Greek means "pure, fresh air" or "clear sky". In Greek mythology, ether was thought to be the real essence that the gods breathed, filling the whole space where they lived, analogous to the air breathed by mortals. The derived word *ether* (also spelled aether) is defined in the *Oxford English Dictionary* as a "hypothetical medium, supposed to fill space, by means of vibrations in which light and other forms of radiation are transmitted".

In his manuscript *On the Heavens* Aristotle introduced a new element to the classification of the elements of Ionian philosophy (earth, water, fire, and air or dry, wet, hot, and cold); he noticed that the four classical elements were subject to change and naturally moved in a linear pattern. The new element however, located in the celestial regions of heavenly bodies, moved circularly and had none of the qualities the terrestrial classical elements showed. It was not characterised by any typical empirical quality, it was neither hot nor cold, neither wet nor dry. Similarly, Plato alludes to ether as the most translucent kind of substance but then he adopted the older system of four elements.

Following Greeks, the medieval concept of the universe was based on a mixture of five elements, earth, fire, air, water and ether and a subdivision of space into concentric spheres. The innermost spheres are the terrestrial spheres, while the outer are celestial ones made of ether and contain the stars.

Up until the latter part of the last century the ether theory was an established scientific fact. The Michelson-Morley experiment, carried out in 1881, concluded that there was no Earth motion relative to the ether (see *Encyclopedia Britannica*, Vol. 8, p. 98). This experiment, which was a one off experiment, discredited the ether theory and caused its rejection in favour of the theory that space is a vacuum and air is merely a chemical composition of Oxygen and Nitrogen plus other minor constituents.

With the 18th century developments of physics, mathematical models known as "ether theories" made use of a similar concept for the explanation of the propagation of electromagnetic and gravitational forces through space.

As early as the 1670s, Newton used the idea of ether to help match observations to the rules of gravitation he developed. All these ether theories are considered to be scientifically obsolete now, as the special relativity and edited version of Maxwell equations do not require its presence for the transmission of basic physical forces. However, Einstein himself as well as Maxwell theories can be interpreted as if the empty space between objects had its own physical properties.

The "orthodox" scientists accept that there is no ether because of the Michelson and Morley experiment. More informed researchers might indeed add the results of a further experiment performed by Trouton and Noble in

1903. Yet the ether theory was dismantled way too easy. An experiment has been performed by E. W. Silvertooth, a retired optical physics professor, who tried to recreate the Michelson-Morley experiment, this time avoiding the retro-reflections using a transparent photodetector in linear translational motion relative to the optical light source to scan along a beam set up by interfering rays coming from opposite directions. He discovered that the whole apparatus was in motion through space in the direction of Constellation Leo at nearly 400 km/s or 248 miles/s. This is what both Michelson and Morley, and Trouton and Noble were trying to find in their experiments.

So why did the Michelson-Morley experiment fail? Well, its erroneous conclusions are mainly a result of the imprecise semi-reflective mirror surfaces causing an optical drag alongside the motion of the apparatus. That error affected the calculated speed of light relative to the Earth in standing wave components along the beam. Michelson and Morley could not know about standing waves properties because these phenomena were only discovered much later by Wiener. The Trouton-Noble experiment did not work as it assumed the presence of Lorentz forces in a circuital motion, a profound misinterpretation any physics undergraduate would understand.

In the 60s a quite extrovert Russian scientists called dr. Nikolai Kozyrev scientifically proved that unseen energies indeed exist all around us. Unfortunate development of his research was pointed towards the creation of Soviet military defence applications which exploited principles of "torsion fields" (e.g., etheric energy) to redirect long-range missiles. Kozyrev's work, which indeed confirms both Reich's research and current empirical experiences with Orgone, was secreted until the fall of the USSR in 1991.

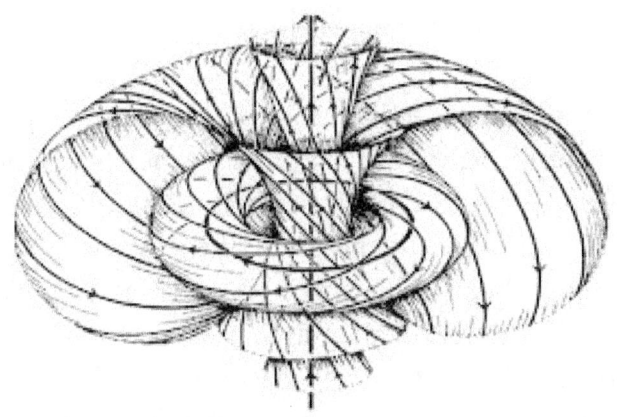

Pictorial representation of Kozyrev torsion field

Providentially, the latest theoretical studies show the correlations between the ether and dark energy and many renowned scientific institutions have found experimental confirmation that a possible effect of etheric drift is detectable and has been detected. In spite of the early ether models being supplanted by general relativity, incidentally a few physicists have attempted to reintroduce the idea of ether trying to address apparent theoretical and practical unexplained phenomena (simply put "lacks") in current physical models. One proposed model of dark energy has been named *quintessence* by its proponents, to pay tribute to the old way the ether was called. This idea relates to the hypothesis of a dark energy formulated as a clarification of perceptions of an accelerating universe. It has additionally been known as the fifth fundamental force of physics, aside to the electromagnetic, gravitational and strong and weak nuclear forces.

The use of ether to depict movement of light was popular during the 17th and 18th century, including a hypothesis proposed by Johann II Bernoulli, who was rewarded in 1736 with the prize of the French Academy. In his hypothesis, all space is permeated by ether containing "excessively small whirlpools". These whirlpools allow ether to have a specific elasticity, transmitting vibrations from the corpuscular parcels of light as they travel through the medium. This hypothesis named "luminiferous ether" would impact the wave hypothesis of light proposed by Christiaan Huygens, in which light propagated as longitudinal waves by means of an "omnipresent, transparent, perfectly elastic medium having zero density, called ether". At the time, it was felt that in order for light to go through a vacuum, there must have been a medium filling the void through which it could propagate, as sound through air or ripples through water in a pool.

Afterwards, when the idea of light and EM waves in general, were clarified as transversal-vectorial rather than longitudinal-scalar, Huygens' hypothesis was supplanted by consequent speculations which dismissed the presence and need of ether to clarify the various optical phenomena. These hypotheses were supported by the results of the Michelson-Morley experiment in which evidence for the motion of ether was conclusively absent. The results of this experiment influenced many physicists and modern theories.

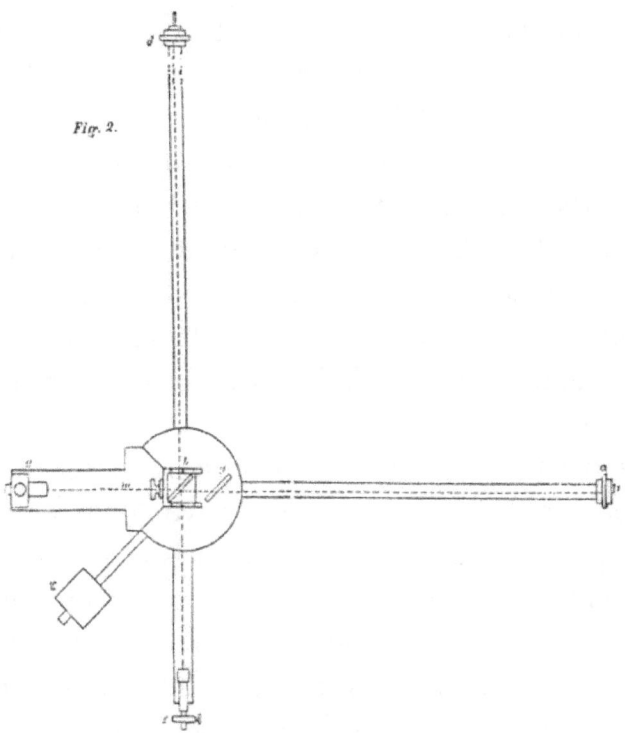

Light travels faster on the arm parallel to the movement of the ether, if ether exists.

Ether has been used in different gravitational theories as a medium to help clarify the causes of gravitational forces. It was even mentioned in one of Sir Isaac Newton's early versions of his hypotheses on the nature of attractive energy, as stated in his famous book *Philosophiæ Naturalis Principia Mathematica* (usually referred to as *Principia*). He based the entire description of planetary motion on a theoretical law of dynamic interactions between distant bodies by presenting a system of propagation through a mediating substance. He called this medium ether and strongly disagreed on the possibility of having a force acting over far off bodies without any direct connection, i.e. excluding the existence of vacuum. In his ether model, Newton depicts it as a medium that flows consistently downwards toward the Earth's surface and is partially absorbed and partially diffused. This theory implies the existence of ether densities gradients, the real origins of ether motion. Ether would be more dense nearby heavy objects and more rarefied when away from massive elements.

As particles of denser ether come closer with the rarefied ether, an attraction force would be exerted to bring rare ether particles back to the dense ones, much like condensation work (cooling vapours of water are attracted back to each other to form waterdrops). An interesting side note is the absolute consistency of Newton's interpretation of gravity with the explanations given in ancient Vedic scripts. Although Newton eventually changed his theory of gravitation to exclude ether to praise other scientists, his starting point for the modern understanding and explanation of gravity came from his original ether model on gravitation.

Tesla too was fascinated by the huge potential of the "enormous energy reservoir" represented by ether.
He considered its radiant energy very powerful and built machines to prove the possibility to harvest free-energy from it. The way of employing radiant energy is to convert waves to radiant energy. Waves in Tesla's radiant energy are very different to today's interpretation, as they coexist in two different states, both immovable waves in ether and longitudinal waves in radiant energy. Tesla's invention of radiant energy is able to harness energy from the air and possibly from anywhere in the universe. However, what is published about the radiation energy generator is mainly in the form of a model, the building scheme and specifications of the patent are still a mystery.

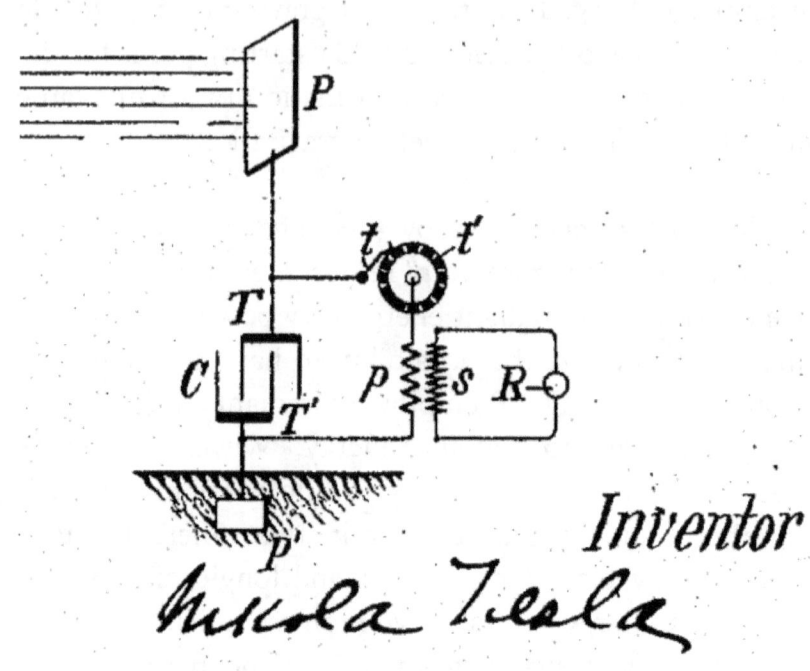

Tesla's radiant energy measurement device

None of his ideas regarding free-energy was taken seriously and this whole part of his invaluable legacy was totally discarded.

Robert B. Laughlin, Nobel Laureate in physics, professor emeritus at Stanford University, had this to say about ether in contemporary theoretical physics.
"It is ironic that Einstein's most creative work, the general theory of relativity, should boil down to conceptualising space as a medium when his original premise [in special relativity] was that no such medium

existed [...] The word 'ether' has extremely negative connotations in theoretical physics because of its past association with opposition to relativity. This is unfortunate because, stripped of these connotations, it rather nicely captures the way most physicists actually think about the vacuum...Relativity actually says nothing about the existence or nonexistence of matter pervading the universe, only that any such matter must have relativistic symmetry. [...] It turns out that such matter exists. About the time relativity was becoming accepted, studies of radioactivity began showing that the empty vacuum of space had spectroscopic structure similar to that of ordinary quantum solids and fluids. Subsequent studies with large particle accelerators have now led us to understand that space is more like a piece of window glass than ideal Newtonian emptiness. It is filled with 'stuff' that is normally transparent but can be made visible by hitting it sufficiently hard to knock out a part. The modern concept of the vacuum of space, confirmed every day by experiment, is a relativistic ether. But we do not call it this because it is taboo".

Bestowing the philosophical point of view of Einstein, Dirac, Bell, de Broglie, Maxwell, Newton and other great physicists of the time, acknowledged that there might be a medium with physical properties filling empty space, a kind of ether, enabling natural physical processes to happen.

In 1984 or 1985 in his first publication, a young Albert Einstein wrote: "The velocity of a wave is proportional to the square root of the elastic forces which cause [its] propagation, and inversely proportional to the mass of the ether moved by these forces". Later in 1920 he added: "We may say that according to the general theory of relativity space is endowed with physical qualities; in this sense, therefore, there exists an Ether. According to the general theory of relativity space without Ether is unthinkable; for in such space there not only would be no propagation of light, but also no possibility of existence for standards of space and time (measuring rods and clocks), nor therefore any space-time intervals in the physical sense. But this Ether may not be thought of as endowed with the quality characteristic of ponderable media, as consisting of parts which may be tracked through time. The idea of motion may not be applied to it". (Einstein, *Ether and the Theory of Relativity*, 1920)

A colleague of his, Paul Dirac wrote in 1951: "Physical knowledge has advanced much since 1905, notably by the arrival of quantum mechanics, and the situation [about the scientific plausibility of Ether] has again changed. If one examines the question in the light of present-day knowledge, one finds that the Ether is no longer ruled out by relativity, and good reasons can now be advanced for postulating an Ether... We have now the velocity at all points of space-time, playing a fundamental part in electrodynamics. It is natural to regard it as the velocity of some real physical thing. Thus with the new theory of electrodynamics [vacuum filled with virtual particles] we are rather forced to have an Ether".

John Bell in 1986 in *The Ghost in the Atom* reportedly said that a well-formed ether theory might help resolve the Einstein-Rosen paradox by creating a convenient reference frame in which signals go faster than the speed of light in vacuum. In that case, Lorentz contraction would be perfectly coherent, not inconsistent with relativity, and could result in an ether theory perfectly in line with the Michelson-Morley experiment. Furthermore, Bell suggests the ether was historically wrongly rejected on purely philosophical grounds: "what is unobservable does not exist". Besides the arguments based on his interpretation of quantum mechanics, he advocates for resurrecting the ether because it is a useful pedagogical device, since many problems are solved more easily by imagining the existence of ether.

More recently, the Romanian physicist Ioan-Iovitz Popescu in 1982 wrote that the ether is "a form of existence of the matter, but it differs qualitatively from the common (atomic and molecular) substance or radiation (photons)". It is a fluid "governed by the principle of inertia and its presence produces a modification of the space-time geometry". Popescu's theory speculates a finite Universe "filled with some particles of exceedingly small mass, traveling chaotically at speed of light and material bodies made up of such particles called etherons". (Duursma, *Etherons as predicted by Ioan-Iovitz Popescu in 1982*, 2015)

ZERO POINT ENERGY

Quantum physics has come to the conclusion that matter, on a subatomic scale, is energy. Mass and energy are intrinsically the same quantity. The nucleus of an atom, its material core, is infinitesimally small in comparison with the size of the atom. Interestingly, the atom is almost all made of empty space, with a highly dense nucleus and some tiny electrons revolving around. If the entire space within an atom were to be filled with particles the size of its nucleus, one million billion of such particles would be required to do so. Thus, an atom is essentially empty. Empty space is essentially embroidered in the nature of things. Is such empty space just vacuum? That is the question.

Today, scientists are more and more convinced that there is a kind of fundamental non-electromagnetic energy pervading space and interacting with matter. Because of its non-conformity to the laws of thermodynamics, it has been referred to as free energy, vacuum energy, scalar wave or quantum energy. A universal, unseen energy medium they call "dark matter", "vacuum flux" or "zero-point energy" with essentially the same characteristics as Reich's Orgone. In Eastern cultures, this energy has been known as life force, *Prana* or *Chi*, etc. In this book, we will be using Aristotle's term *Ether* or etheric energy to describe this subtle energy.

The well-known ether theory developed during the 19th and 20th centuries revolves around the idea of ether as a static medium. However, ether is a highly dynamic vibrating energy in constant motion which always follows a vortex and spiral path. All manner of different vortex sizes and shapes exist and etheric energy moves in a vortex. In 1874, John Worrel Keely invented a machine to be able to harvest power from etheric energy and convert it into mechanical work. He never disclosed the inner workings, just that it revolved around a water jet electrolysed by etheric energy. This invention was patented as the Keely engine and was the reason behind a long court trial between Keely himself and several investors, who were mainly interested in buying his building plans for a ridiculously cheap rate. He wrote in 1893: "there is no dividing of matter and force into two distinct terms, as they both are one. Force is liberated matter. Matter is force in bondage".

His theories and inventions were also opposed by powerful organisations with vested interests in promoting the steam engine, to which he eventually succumbed. His understanding of life energy was beyond that of contemporary scientists. Being far ahead of his time, akin to dr. Wilhelm Reich, he was ostracised for his revolutionary ideas. Despite the problems Keely encountered, other people such as Nikola Tesla, Viktor Schauberger and Wilhelm Reich have kept investigating and have scientifically proven that a form of etheric energy, although invisible and weakly interacting with ordinary matter, is very real. It is the substance astrophysicists had to bring in their equations to justify the existence of galaxies; however, to avoid persecution by the secular authorities, they granted it the designation of "dark matter".

WILHELM REICH

Wilhelm Reich in his lab

Wilhelm Reich (1897 - 1957) was an Austrian medical doctor and psychoanalyst who theorised the existence of a new form of energy named Orgone and became one of the founding members of modern psychiatry. He entered the University of Vienna therapeutic school in 1918 where he quickly became one of the main followers of Sigmund Freud teachings. Here he rapidly turned out to be exceptionally regarded as a therapist and psychoanalyst. Strongly bond to Freud ideas, he got Freud's help to start his career accepting patients since 1920, at just twenty-three years, when Reich was already acknowledged as a professional therapist from the Vienna Psychoanalytic Association. Only a couple of years after, in 1922, he started his private practice and enrolled to be part of Freud's Psychoanalytic Polyclinic.

His constant comparison with Freud was source of distress and caused an abrupt scission from the psychoanalytic foundation. It was expected, to some degree, due to Reich's interest for issues outside the direct control of psychology. In fact, by that time dr. Reich had turned into a questionable figure, having started to experiment with Orgone energy and (meta)physical matters.

Wilhelm Reich was an astoundingly productive essayist, with fifteen books and a copious number of scientific articles and publications. His *Character Analysis*, distributed in 1933, was pivotal, as it hypothesised that an individual's general character, as opposed to just their external manifestations, ought to be viewed as when diagnosing and dissecting anxiety. Critically, it was *Character Analysis* that presented the hypothesis of "resignation" that Reich considered to be the main factor that blocked psychosexual energy causing an internal energetic depletion in the individual as well as physical illnesses. He would keep on holding to his thesis promoting a rehabilitation of the sexual sphere from the coercive laws of his contemporary society.

In 1939, just before the beginning of World War II, he left Europe for Norway and then for the United States. In 1940, having demonstrated the presence of Orgone energy in the air, he assembled the first "Orgone Energy Accumulator" (ORAC), a box-like structure intended to probe environmental orgonic energy fields. He likewise conducted preliminary studies of the Orgone accumulator on patients experiencing a vast range of sicknesses. During those years Reich kept in touch with the famous physicist Albert Einstein, exposing him his discoveries of the Orgone energy and the ORAC. He visited the scientist in 1941 in Princeton, where they talked for a long time and tested the device in person.

In 1947 the Food and Drug Administration (FDA) began to investigate on Reich's studies and claims. Reich's work was found to have socialist incited ideas, and that such theories could resemble a dangerous provocative praise of communism during a specific historical time where Cold War was raging. Others thought that his hypotheses regarding a form of life energy, and particularly the relevance of the orgasm and free sex as a psychological relief were misleading and unacceptable and prompted the government to take action.

In May 1956, Wilhelm Reich was captured for abusing a court directive. Without Reich's approval and going against his specific requests, a research fellow deliberately sold and shipped Orgone devices over state borders. This fact negatively impacted on the trial. Reich was not guarding his untested devices and was lastly uniquely charged for having neglected to comply with the court's order. He was condemned to two years in jail. The FDA destroyed all Orgone accumulators and burned, in four separate events, his books and writings. Wilhelm Reich died on November 3, 1957 in a government prison in Lewisburg, Pennsylvania.

That's what the FBI says about him: "This German immigrant described himself as the Associate Professor of Medical Psychology, Director of the Orgone Institute, President and research physician of the Wilhelm Reich Foundation, and discoverer of biological or life energy. A 1940 security investigation was begun to determine the extent of Reich's communist commitments. In 1947, a security investigation concluded that neither the Orgone Project nor any of its staff were engaged in subversive activities or were in violation of any statue within the jurisdiction of the FBI. In 1954 the U.S. Attorney General filed a complaint seeking permanent injunction to prevent interstate shipment of devices and literature distributed by dr. Reich's group. That same year, dr. Reich was arrested for a Contempt of Court for violation of the Attorney General's injunction".

THE DISCOVERY OF ORGONITE

The authorities didn't succeed in burying Reich's ideas and in the '80s a young Austrian scientist, Karl Hans Welz, started to research prior evidences of life energy in history. He went on to analyse a vast amount of literature from Mesmer to Tesla works, finding several occurrences of an inexplicable form of energy most previous scientists and philosophers had just ignored. Welz discovered that orgonic energy can be projected at a long distance using structural matter bridges so he decided to develop a mathematical model for such phenomenon. Now, he turned theory into practice when building the first Orgone Generator, a device able to create Positive Orgone from negative one using the so-called DOR-POR reaction or concurrent (de)polarisation and bions carrier charge inversion. Welz started selling the device, an EPG 2000, since 1992; this first iteration was using alternating layers of steel wool and fibreglass in a cylindrical shape. Further iterations showed how englobing metallic particles inside resin was a better approach to increase the overall device efficiency. A later yet powerful addition was made at the beginning of the 21st century by Don and Carol Croft who added a quartz crystal to the metal-resin mix; the crystal would resonate to the DOR frequencies and increase its energy causing a spontaneous transition into POR.

Actually, historically Orgonite was already present in Franz Bardon 1956 book "The practice of magical evocation" where he mentioned a device capable of storing ether energy, a "fluid condenser" made of layers of metal and plastic (the same of Welz' Orgonite).

The way Orgonite works can be summed up as the consequence of multiple effects combining together. First, thanks to the well-known layering of the ORAC, Orgone is captured from the environment and brought inside the pyramid. At the same time, the copper spiral resonates with electromagnetic waves as well as DOR causing a magnetostrictive effect on the piezoelectric quartz crystal. Magnetostriction is the property of metallic materials to change shape due to electromagnetic fields while piezoelectricity is the accumulation of electrical charge in materials when they are compressed or enlarged. The copper thread will shrink and expand causing a piezoelectric effect on the crystal, an accumulation of charge on the crystal termination points using the power of tips where the electromagnetic field becomes stronger. This is where the DOR-POR reaction happens and where dangerous electromagnetic fields are absorbed by the crystal and converted in electrostatic energy needed for the aforementioned reaction to happen. Positive Orgone is then released in the environment from the pyramid sides.

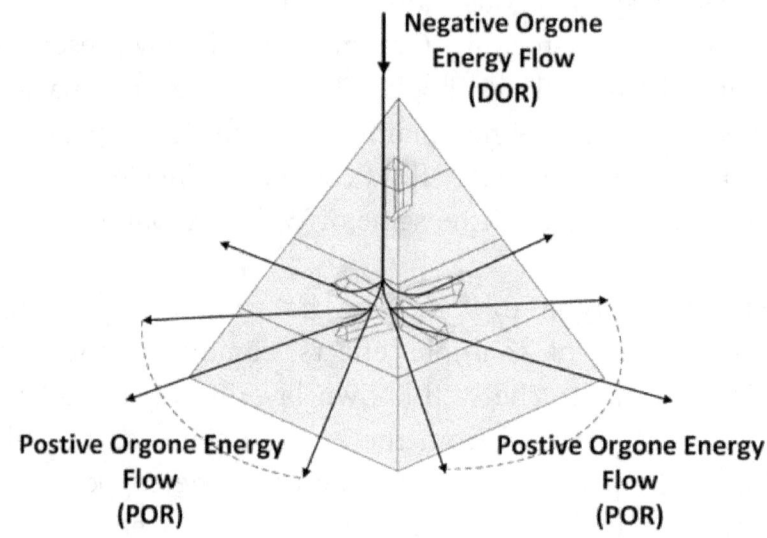

Scheme of the way an Orgonite works

EXPERIMENTAL EVIDENCE OF ORGONE

DIRECT OBSERVATIONS

In *The Cancer Biopathy*, chapter IV, section 5, Reich goes on to talk about three experiments that can demonstrate the inner pulsation of Orgone energy and its direct linked to the carrier charge of the bions.

The first one is the Orgone pendulum, in his words: "A metal sphere of iron or steel, about 4 to 6cm. in diameter, is placed on a stable surface, a solid table for instance. A much smaller sphere, about 1cm. in diameter, is suspended pendulum-wise at about 0.5cm. from the equator of the larger sphere. For definite reasons the length of the pendulum thread should be exactly 16 cm. [...] My experience is that the best results are obtained by making the pendulum sphere out of a mixture of soil and iron filings (i.e. a combination of organic and metallic material) moulded together in water and then put into an extremely thin-walled glass sphere. The bigger sphere and the pendulum sphere are then covered with a cellulose cover to protect them against air currents". The pendulum would then oscillate towards the metallic sphere when there was good weather and stop in case of bad weather or if the viewer was in a bad, depressive mood. Strong emotions can highly affect the orgonic energy field causing a depolarisation of POR into DOR.

The second one is the observation of ripples in the air using a telescope. Reich set up a 9cm or 3½" telescope on one shore of Lake Mooselookmeguntic in Maine during two consecutive summers, and, during the daytime, peered through it toward the opposite shore of the lake 4-8 miles away: "When the telescope is pointed south, it is possible to observe against the background of the opposite shore of the lake, at a magnification of only 60x, a wavy, pulsating movement travelling, with few exceptions, always from west to east. The west-east movement is constant, whether the lake is rough or smooth, whether or not there is wind, and whether the wind is from west to east or south to north, strong or weak. The further the telescope is turned toward the west or east the more difficult it is to see the movement. It can no longer be seen when the telescope is trained due west or east". (Reich, *The Cancer Biopathy*, 1973).

He noticed a strange motion of air masses from west to east causing a trembling in the air and explained the phenomenon as a cluster of atmospheric Orgone, pulsating due to external distress.

Last observation is traces of ionised bions moving in a spiral pattern across his field of vision. In *The Cancer Biopathy*, chapter IV, section 1, he wrote: "When we were children the light phenomena we saw with our eyes shut were a constant source of fascination. Small dots, blue-violet in color, would appear from nowhere, floating back and forth slowly, changing their course with every movement of the eyes. They floated quite slowly in gentle curves, looping periodically into spirals, in a path somewhat as follows [see image]".

Just like neutrinos coming in huge amount from the Sun like a constant shower on Earth albeit very rarely interacting with matter, the same happens with Orgone bions passing through the environment like neutrinos, yet greatly reacting with organic matter.

THE ORAC AND THE EFFECTS OF ORGONE ON LIVING BEINGS

In 1942 dr. Reich revealed his trials on the treatment of rats with spontaneous mammary tumours with an *Orgone Energy Accumulator* (ORAC) device. The ORAC was one of the direct results of Reich's investigations started years sooner into the real origin of what he defined "the cancer biopathy" type of affliction. Through clinical trials, Reich had discovered that a cancerous disease was definitely more than the mass of rapidly developing mutated cells known as a tumour, instead it was a multifactorial disease rooted in the living being as a whole: cancer was yet the most noticeable physical appearance of the disease processing from one stage to another. Reich noticed all malignant cancer patients at a profound mental level show a regress "resignation" emotional struggle, that is, abandoning living as an animal embracing life in all its forms, instead forcefully blocking all the unacceptable parts of their essence (mainly sexuality and love) in the way it was intended to be. Deep emotional regret and resignation have got the capacity to induce via the psyche some visible biophysical changes in the body, namely the constant alteration of the "bio-vitality" energy fields within the organism. This claim was supported through evident hypersympatheticotonia, a physical body condition induced by great hyper stimulation of the sympathetic nervous system and characterised by lightening of the skin, goosebumps, skin vascular spasms, increased pilomotor activity (movement of hairs), and abnormally high blood pressure. Furthermore, there was a general depletion of bioenergy levels in the organisms. All previous research on cancer used biopsy and tissue material as living sample, thus neglecting the interference and the effects on the living being as a whole unique complex interconnected system, instead relying on the archaic method of *divide et impera* very dear to modern allopathic medicine.

It seemed obvious how cancer, as well as many other diseases with no direct obvious source (like a wound, a toxin or a virus), could be the result of a concatenation of causes where the psychological part played a major role. Not only neglecting the emotional side of the person could cause a nervous breakdown with serious health effects, but also the intangible change of environmental conditions affect bioenergetic fields in the organism and result in a plethora of symptoms with no apparent explanation.

When researching with different materials and photographic films, dr. Reich said to have noticed some glowing blue light flashes coming from deteriorating organic samples (like over-ripen fruit), which he managed to see impressed on photosensitive films. According to radiologists at Oslo hospital, it could not be any known form of known radiation: it was the hint that suggested Reich the existence of an unknown type of energy instilled in everything alive, something he later called *Orgone*. In fact, as it was very visible with oncologic patients, repressing the vital and sexual part of the individual actually meant depleting the organism reservoirs of living force.

Now that he had an explanation for the nature of such energy found in people, animals and plants, he needed to investigate the connections among creatures and Orgone.

Since Wilhelm Reich had started his researches looking at bioelectricity in living beings and its structure, he developed a whole new theory where Orgone energy was made of elemental particles called bions. Their distinguishing feature is the blue ionisation sometimes related to their presence.

Reich explored the nature of these new particles in his 1937 work entitled *The Bion Experiments on the Origin of Life* where he states that bions are microscopic structures half-way to the elemental particles of physics and living cells of organisms.

Bions internal structure is composed of smaller vesicles that are visible by looking at the slow disintegration of organic matter in sterile pure water, namely grass and cotton (what he used for testing). Further studies actually identified bions as the vesicles and their aggregation status as excited states of organic matter.

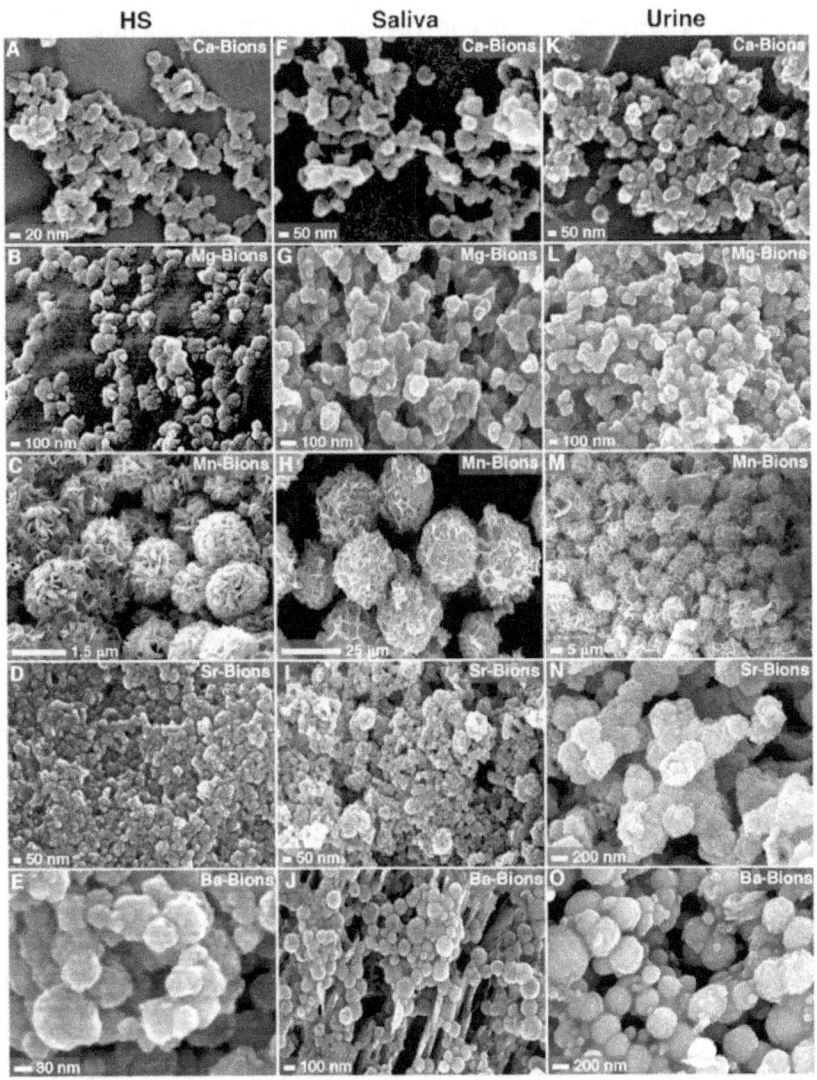

Bions forms found in various human bodily fluids

He was convinced of the presence of such structures in living being that he planned an experiment with amoebae. That's what he said about it:

"In the course of about fifteen years of clinical work, I came to recognise a formula for the function of the orgasm which was verified in subsequent experiments [described in *The Bioelectrical Investigation of Sexuality and Anxiety*]. In vegetative life there is a process through which mechanical filling, or tension, leads to a build-up of electric charge; this is followed by electrical discharge, which, in turn, culminates in mechanical relaxation". (Reich, *The Bion Experiments*)

Such subsequence of events

```
mechanical tension ->
bioenergetic charge ->
bioenergetic discharge +
mechanical relaxation
```

is deeply linked to Reich's vision of sexual energy (borrowed from Freud's *libido*). The explanation is still left uncleared as he didn't finish the article about it, what can be deducted is that bions act as the origin of motion and mechanical work acting like the electromotive force moving an electrical motor.

Reich saw that bions can easily interact with organic beings, from animals to plants, inducing a huge plethora of visible physical effects, mainly related to the general wellbeing. As highly energetic photons can cause cancer while low energetic ones can heat us, same goes for bions: they can acquire a negative charge and cause negative effects on organisms or a positive one and provide health benefits. "Decrease" and "Increase" of potential could be interpreted in terms of functional physics simply as a change in the form of the atmospheric energy from the fog-like (unexcited, low) to the pointed (excited, high) state of existence and vice versa. Only the pointed state can be detected by the Geiger-Müeller counter.

One remarkable feature of Orgone energy is its ability to spontaneously luminate through air ionisation. Reich reported that in the obscured laboratory within the research centre, he (and others) could perceive dark blue vapours wrapping organic materials and tiny, transient, scattered particles flowing like "beams" in the air. So as to more readily test these breakthroughs, Reich fabricated an enclosed chamber made of a glass front and several dividers of different materials layered in alternating order. The metal surface would reflect the radiation from the encased bions emitter and the wood would assimilate it, the result being a collimation and amplification of the radiation inside the internal area.

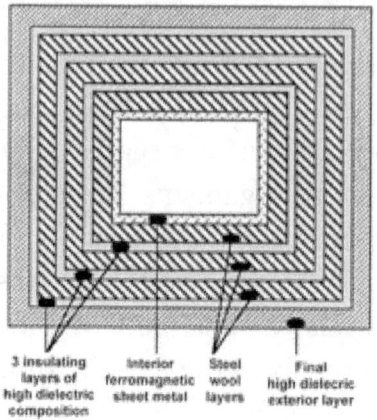

3 insulating
layers of
high dielectric
composition

Interior
ferromagnetic
sheet metal

Steel
wool
layers

Final
high dielectric
exterior layer

Reich revealed he was, in fact, better able to see the typical Orgone sparks, however incredibly they persisted even when taking the source of bions out of the chamber. Wiping the inner surfaces, airing out, and adding another layer without the organic part did not attenuate the phenomenon. Observations made it evident how such energy could be collected from the environment and concentrated in a small area using his ORAC. The new type of energy is confined and attracted by metallic materials and consumed by non-metals.

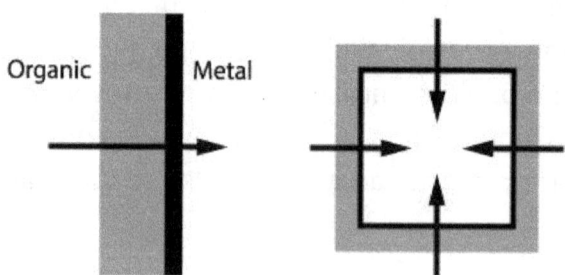

Organic   Metal

This finding was followed by several tests and examinations, one of them being measuring the temperatures inside the accumulator compared to temperatures of the outside air or a reasonable control chamber. Reich found out the temperature inside the ORAC, especially its upper parts, was higher than the rest of the ORAC and on overall the inner chamber was warmer than the air around the ORAC over a reasonable value. Materials were also subject to tests. For the metal layers iron and steel wool appeared to be, while various plastics and fabrics seemed to be well-suited for the non-metallic layers. When working with plants or living creatures, including people, it is significant to notice that the ORAC must be slightly bigger than the subject; the working principle is a shared excitation and relaxation between the orgonic energy charge of the individual (or creature or plant) and the accumulated Orgone inside the ORAC. There must be adequate air gaps in the dividers or entryways and around the living being in order to reduce detrimental energetic turbulences.

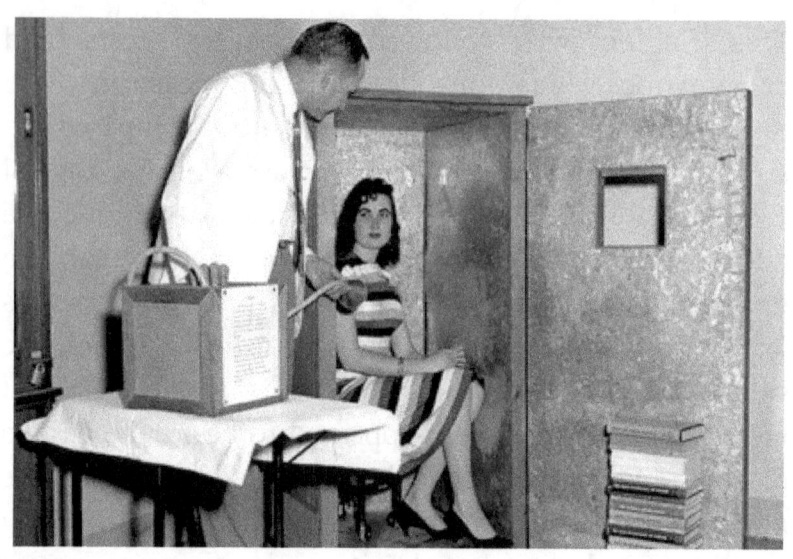

A patient inside the ORAC

Reich issued numerous articles and a few books on the practical working design of the ORAC and derived devices that allowed the analysis of Orgone energy and its charge. Students of Orgonomy have collected a considerable amount of his texts and lab notes, all of them have been distributed through Reich's Orgone Energy Bulletins, the American College of Orgonomy's Journal of Orgonomy, the Institute for Orgonomic Science's Annals, The Orgone Biophysical Research Laboratory and many others.

While Reich and others had found many interesting uses of the ORAC, the main test bed was malignant cancer in mice. Before starting to see the impacts of the ORAC on tumours, Reich had considered the possibility of injecting emitters of highly energetic bions acquired from sea (sand parcel, *SAPA bions*) into mice with mammary tumours. Some strikingly anomalous results were noticed: for instance, a fast reduction of the tumour mass at times and a consequent extension in the life expectancy of treated subjects. The SAPA bions evidently reinvigorated the organism's bioenergy reservoirs and instilled vitality into the test subject immune system, which at that point assaulted the malignant cells. A series of tests including the administration of a serum enriched with SAPA bions to an animal, demonstrated an evident damaging effect on the animal's cancerous cells; however, such strategy was not as powerful as the direct exposure to a high source of pure bions. Looking at and comparing samples of liver tissue of treated and untreated animals, Reich noticed that a breakdown of the cancer into smaller carcinogenic sub-cells was the common denominator to health recovery.

With the invention of the ORAC, Reich was finally able to see the impacts of irradiating mice with malign mammary tumours inside an Orgone accumulator. Accumulators used in the tests consisted of an internal mass of pure iron and an external layer(s) of steel fleece and a fibrous material, for example, cotton, wool, or celotex as an external auxiliary help. The mice were selected from a strain, the "Rockford", that can spontaneously develop malign mammary tumours.

The ORAC was split into sectors so that every compartment could keep a couple of mice inside. Treatment started a couple of weeks after the discovery of the cancer and kept going for 1/2 an hour, day by day, until total regression. Reich carefully reported the day of the death of treated and untreated control mice. The outcomes were shocking: the treated mice lived much longer than the control mice, most of the times showing a complete health recovery. A specific element in the malignant cell research is Reich's perception that the blood of an oncogenic mice did not create direct antibodies against disease cells. The reason for that, he disclosed, is due to the "Orgone-blocking" effect of the damaged blood cells in cancer.

The entire human body creates a strong orgonic energy field, frequently called "vitality" or "aura". A visible example is the astronaut walking on the Moon surface surrounded by a glowing blue light. Typically, we replace the consumed Orgone by eating and drinking, which brings in bions in the body via direct absorption (for water) or food processing in the digestive system; additionally, breathing takes in Orgone energy directly from the environment into the lungs and blood; likewise, the skin can attract orgonic energy, mainly when exposed to sunlight, which Reich felt was a noteworthy source of Orgone dispersed in the environment. All these sources refuel the vitality in the organism and rebalance the internal energetic potentials that drive many biological systems fundamental for the body homeostasis, the maintenance of a low entropy status related to a healthy condition. Atmospheric physical effects were seen happening outside the ORAC, demonstrating a strong abnormality and electroscopic instability, showing the chamber is gathering an undeniable enormous amount of vital energy. Similarly, such effects remember what dr. Puthoff experienced when looking at high-vacuum tubes and the "zero-point vacuum fluctuation".

In 1941, Reich gave an Orgone accumulator box (with thermometers set up in strategic spots) to Albert Einstein. To his surprise, Einstein immediately found that the temperature inside the reaction chamber was higher than the temperature of the surrounding environment. Later on, Einstein simply recalled that the difference in temperature was due to air convection despite the controlled lab setting aimed at eradicating such cause of measurement errors.

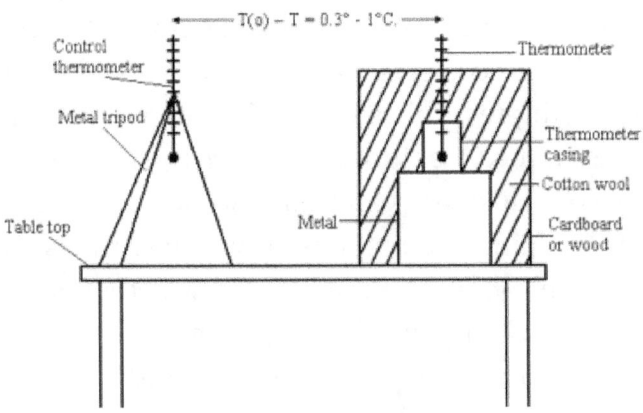

Experimental test to compare temperature inside an ORAC and outside

THE T-BACILLI

At the point when Reich was focusing on the bions study, he saw that blood cells, when fixed and decolorated, can deteriorate into bions. Reich thought that they were the transporters of Orgone energy just as they do with oxygen through all the body. He further saw that when individuals were sick, the bioenergetic field around the blood cells was smaller, and their decay into bions was faster, sometimes actually showing the presence of smaller bacilli (T-bacilli), a symptom of a dangerous condition in the body. Named T from the German word *Tod* (death), those bacilli measure 0.2-0.3 microns and, when cultured, smell like a rotten corpse. Injected into a mouse in large doses, T-bacilli can kill in less than 24 hours.

In chapter II, section 1 of *The Cancer Biopathy*, Reich described the XX Experiment, investigating the presence of corpuscular entities inside organic matter. Everything started with a "coal bion preparation". It was obtained by autoclaving test tubes containing 50% organic matter and 50% 0.1n KCl solution in order to sterilise it, then heating coal dust to white-hot incandescence in a gas flame and adding the coal dust to the test tubes while it was still white-hot. The preparation was then subject to analysis. Reich reported: "We examine the stained preparation at a magnification of 3000x, using oil immersion, and find that most of the blue vesicles that previously had every possible form have now become spherical. A new phenomenon is especially striking: alongside the large-sized vesicles, approximately one micron in diameter, there are tiny red bodies which were not visible at a magnification of 300x. The smallest of them are approximately 0.2 micron in length, i.e., only barely visible microscopically. They lie in groups around the larger round, blue vesicles and unstained crystals. They are elongated, and are pointed at one end like miniature lancets". (*The Cancer Biopathy*, ch. II, 1973)

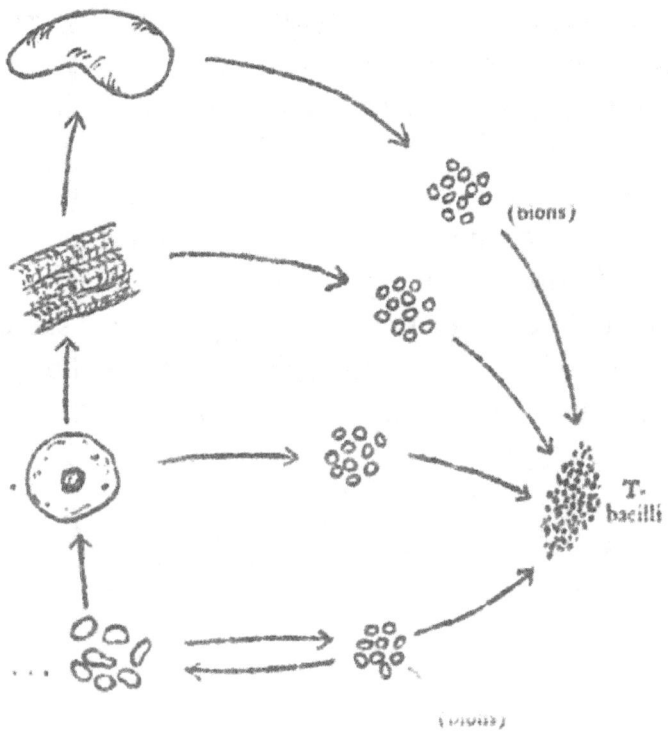

T-bacilli observed in living tissue

THE ORANUR EXPERIMENT

On April 1952, dr. Wilhelm Reich talked about an experiment on the nature and effects of Oranur carried out at the Orgonon laboratory. This examination aimed at finding similarities between nuclear energy (NR) and Orgone energy (OR). dr. Reich needed to understand what happens when Orgone is confronted with ionising radiation. The outcomes are exceptionally critical. By looking at the consequences of the experiment and evaluation of its impacts, namely the consequences for lab mice and measurements of Geiger counters, Reich derived that orgonic energy gets profoundly energised by atomic decay. It acts as a denaturing agent for the carrier charge in the bions turning it to be dangerous (DOR). In a controlled circumstance such an adverse effect can deliver health advantages like a vaccination that stimulates the development of a natural resistance to radiation in the body. Dr. Reich went to obtain two milligrams of Radium-22 and he began the trial.

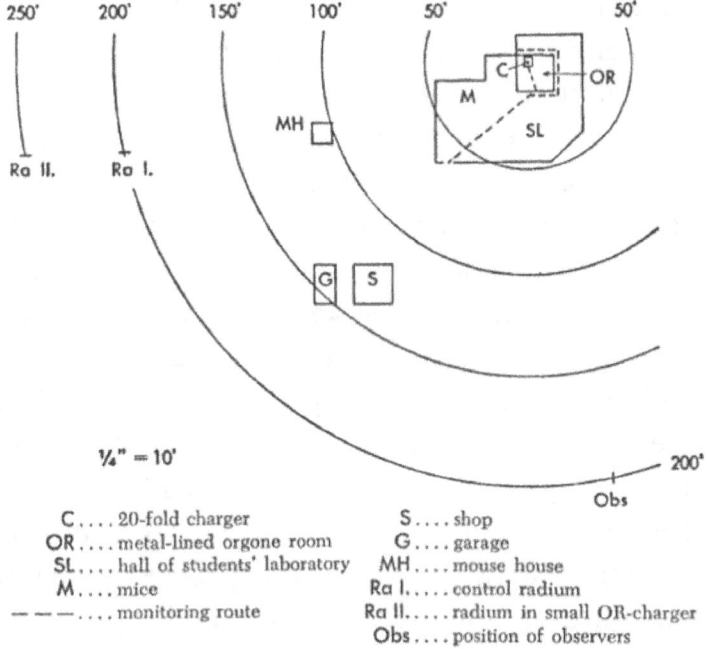

C.... 20-fold charger	S .... shop
OR .... metal-lined orgone room	G .... garage
SL.... hall of students' laboratory	MH .... mouse house
M .... mice	Ra I..... control radium
— — —.... monitoring route	Ra II..... radium in small OR-charger
	Obs .... position of observers

The test consisted in the irradiation of rats by 1 mg of Radium-226. The capsule containing the radioactive isotope was put into a sturdy 20-layer accumulator. This accumulator was additionally placed in the orgone room. Moreover, he kept a group of mice kept in a cage close to the external walls of the orgone room.

The side effects were almost catastrophic: a large number of the employees in the lab experienced sicknesses and the Orgonon and the surrounding research facilities had to be cleared. Considering the troubles and fear that the prosecution of such experiments were bringing to the staff, technicians, and employees, Reich stopped the tests and covered the radionuclides far from the lab.

Myron Sharaf, Reich's colleague and biographer, wrote about the experiment: "Though the official experimentation had stopped around Jan. 20th, the dramatic events connected with its effects were still in evidence till almost the end of Feb., and they are still by no means completely absent. Eva R. almost died on Feb. 19th from sticking her hand in an OEA (Orgone Energy Accumulator) that had been kept shut and in the lower lab constantly. During the same week Mrs. R. and Peter's blood pictures were very bad and they left the lab for a few days and lived in town. Almost everyone who was in the Orgonon lab that week suffered from malaise, (headaches, tiredness, inflamed eyes, and other symptoms). All the accumulators were dismantled around that time, and the metal-lined dark room, too; to this day (March 14th) it is still not possible to have an accumulator (that is, an accumulator at all affected by the experimentation) in the lab without the persons living there suffering from dry throats and lack of air. [...] It is most likely, and even imperative to assume that quite ordinary materials such as rock, metal, and especially material arrangements which have the faculty of accumulating OR energy, continue to be active long after the originally triggering NR has been removed. This resembles induced radioactivity."

Apparently, Orgone did resonate with nuclear radiation and amplified its charge becoming more and more energetic, as well as increasing the ionising radiation energy.

Test mice outside the accumulator were also affected and all died within few weeks from the experiment. It was obvious that all the countermeasures developed by the Atomic Energy Commission were not effective against an Oranur energetic field and that nuclear physics was (and is) still lacking in many parts. The most reasonable way to explain the above phenomenon is falling back on the presence of another physical substance in the environment, that was not new anyway to the thinkers and researchers of the past, pervading the whole Universe, with explicit physical quantities and whose behaviour is highly represented by clear mechanical laws.

This new physical entity likely could be identified with the Orgone energy that Reich discovered permeating everything in nature, including living beings, and all the cosmos. The occurrence of a comparable energy form, in any case, was excluded by Einstein when building up his theories of the natural physical phenomena. He, surely, theorised a perspective on reality without any energetic background. Nevertheless, Einstein at some point came to admit that his models could totally collapse if a new form of energy was found to exist. Clark noted the reaction of Einstein to the results of Miller's experiments performed at Mount Wilson Observatory on the etheric drift, aiming to demonstrate the presence in the Universe of a medium, the ether, through which waves could propagate.

In a letter written to the American physicist Millikan in June 1925, Einstein seems to point out a fundamental doubt about his new theory: "I believe that I have really found the relationship between gravitation and electricity, assuming that the Miller experiments are based on a fundamental error. [...] Otherwise the whole relativity theory collapses like a house of cards."

The last sentence shows how the Nobel laureate was totally aware of the limitations of his theories and that he was becoming more and more concerned about the whole model collapsing.

A schematic way to see the differences between Einstein and Reich's view of matter is condensed in the following picture. The German physicist sees air as a unique component. Sure there are several chemical elements dissolved, nitrogen, oxygen, carbon dioxide, noble gasses but no energy is present and perceivable. Reich instead imagined that to have a complete model or nature of air (in this example), in addition to the chemical view of air, we should also consider an energy filling the Universe in a continuous way.

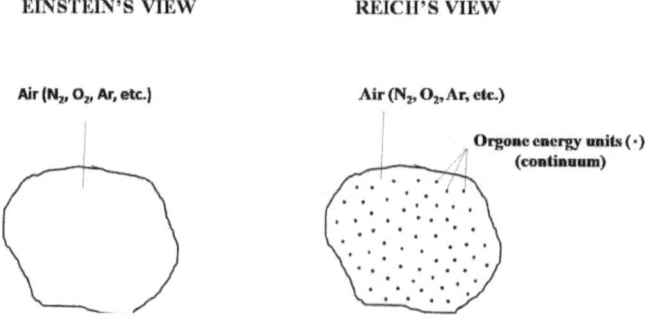

One of the features of these tiny units is their natural wavelength phase shift, a result of some external factors they could be subjected to. Reich found that in normal condition these particles are flowing untouched, with an undulatory movement, and are characterised by energy-enhancing properties. When these units are excited by external factors, they start a chaotic, irregular movement, with a high degree of inhomogeneity. Activation causes of this excited behaviour might be any type of energy, kinetic, electric or electromagnetic, like X-rays and above all radioactivity. Reich considered the charged alpha particles and the photons (γ rays) emitted by radioactive materials as the highest activation agents: "Any kind of mechanical energy is capable of exciting the OR energy to greater activity: simple heat, sparks from an induction coil, friction, etc. The Oranur effect differs from other excitations of OR by its fierceness and its dangerous character".

Under the effect of these physical constraints the free-flowing, undisturbed units start varying their paths and moving in an unpredictable way. At the same time, there is a strict decrease of the life-enhancing properties the particles possess until they lose all the positive qualities and transform themselves in static, deadly, and life-negating units that Reich grouped into the deadly Orgone energy. The latter phenomenon occurs when the triggering agents endure their excitation activity on the units for a period long enough to devoid them of all the vitality they possess. From now on the units behave like deadly aggregates upsetting the surrounding environment and creating a block to the movement of further dynamic units that are trying to flow through. In these conditions the living beings around are exposed to these deadly units and start to agonise and succumb. This event creates either long-term diseases or desert-like conditions, respectively endangering critical organisms functions and extirpating the local flora. Luckily the phenomenon is reversible and, by appropriate conditions and processes, the DOR can be converted back again to positive orgone, promoting the pairing of dynamic, flowing, and life-enhancing orgone units.

A further and equally important characteristic of the dynamic Orgone units is the conversion of energy to matter, an idea already present in Einstein's most famous equation

$$E = mc^2$$

Two or more flowing units can create matter by superimposition, as Reich pointed out: "Atomic energy (nuclear energy) represents cosmic energy which is freed from matter through disintegration of the atom, which is the constituent of the universe in terms of classical and quantum physics. It is energy after matter.

OR energy on the other hand, represents cosmic energy before matter, i.e., energy which has not been caught in or has not been transformed into solid matter. It is universally present, penetrates everything, surrounds, as the so-called OR energy envelope, our planet and most likely all other heavenly bodies (Sun's corona, Saturn's ring, etc.). Cosmic OR energy, moving freely within the living organism, is called bio-energy or organismic OR energy".

A similar test was conducted in 2018 by dr. Southgate with some particular results: the Oranur field did not seem to be affected by environmental conditions and smoke can act as a moderator, lowering the deposited energy. The latter may be the reason why incense is often used in churches and other religious places. He employed 0.26mg of Americium-241, a quantity about 1100 times less than that the one used by Reich in the original Oranur experiment. The radioisotope, stored in its original plastic case, was put for a few weeks either inside an Orgone accumulator, where many smaller orgonic devices were placed, or around it at various distances. The resulting values of the Oranur field were monitored by measuring the radioactivity levels with a Geiger-Muller counter, as well as the values of the Orgone field by an orgonoscope. Dr. Southgate found an anomalous build-up of an Oranur field all around the orgone box with a radius of 5.5-7.3m or 18-24ft confirmed by a comparable intensification of the radioactivity and orgonic energy field readings. He also observed a direct connection between the Oranur field and that the intensity of the Orgone, when exposed to the action of the radioactive isotope, was complemented by higher values of the radioactivity. According to Reich's own experiments, the Oranur effects endured significantly after the removal of the Americium-241 toward the finish of the test and that the cabinet and the Orgone devices contained inside the lab, maintained its Oranur charge for much longer periods.

It should be noted here that in later tests with Oranur, Southgate did produce a field even by using an extremely small amount of radioactive material, with an intensity as low as 0.9mCu. The possibility of the creation of an Oranur field from such a low radioactive source was already predicted by Reich: "Everyone is strictly warned against using an OR accumulator while a NR source of the strength of a millicurie or an X-ray machine is located within the OR energy field. The longer the trigger effect is allowed to act upon the OR atmosphere, and the more OR accumulating devices that are present, the stronger will be the effects. Stronger effects will also result with regular repetition of the trigger action. In a highly concentrated Oranur atmosphere, the presence of even as little as a microgram or less of any NR source for even as little as a few minutes will suffice to produce severe effects".

A possible medical use can be foreseen in controlled exposure of living beings to Oranur: to convey it in the organism, Orgone emitters (resin, steel wool and rocks) were exposed to Oranur inside a glass reaction chamber for several weeks. The resulting sample was found to have an additive 30% to 50% of the original Orgone charge and was placed next to the area to be treated. Results were very interesting and new research is going on for a safe use of such a powerful energy.

Another set of experiments Reich carried out earlier on was mainly aimed at measuring the intensity and occurrence of bioelectricity in the body. He believed that the internal electrical potential rose with pleasure and fell with anxiety, directly connected with Orgone presence. A healthy and mentally satisfied organism would accumulate Orgone from the environment while an unhealthy and unsatisfied one would slowly turn the internal orgonic energy and the surrounding one into DOR.

Reich firmly believed that Freud's *libido* wasn't just a practical concept to describe the individual sexual tension detached from any objective reality, but instead a real form of energy present in all beings and measurable that he identified in Orgone and its carrier particle, the bion. He was persuaded and went on to prove that many electrical phenomena happen in the body and that different mental states can cause an increase or a decrease in the electrical potential depending on their nature, if positive or negative stimuli. That was the basis for a renewed view of life and the symbiosis of all organisms with the surrounding environment.

From 1934 to 1937 during his stay in Norway, Reich completed a series of experiments on human subjects measuring the electrical potential across their skin while making them experience different emotional states, mainly pleasure and anxiety. He thought any electrical changes could be a direct manifestation of the underlying psychological changes happening to the test subject. In order to measure very small electrical potentials on his subjects' skin, he hooked them up to an apparatus made of two electrodes placed on the skin connected to a tube amplifier and an oscilloscope. The experiments were a fundamental step for the comprehension of the Orgone energy effects in the human body.

Such exploration upon the skin potential was already the subject of several experiments dating back to the XIX century by J. Tarchanoff and O. Veraguth, who had recognised that the electric potential within the body would exhibit sharp and abrupt changes in response to direct stimuli. Reich also noticed how an abrasion or a wound can be extremely detrimental to the overall electrical potential measured, corroborating his hypothesis on Orgone flowing out of an unhealthy and mentally negative (feeling pain and suffering, for instance) individual.

The first condition needed for the electrical measurement of sexuality is that skin and mucous membrane surfaces must be undamaged and possess a *resting potential, or basic electrical charge.* If one damages any area of a subject's skin by scratching the epidermis and then applies an electrode to this area, while the other electrode is applied without pressure to various undamaged skin areas, then, when the subject is connected to the electrical circuit of the oscillograph, the light beam deviates from the absolute, otherwise motionless zero line jumping rapidly to a different position. This is because the electrical surface charge of the undamaged skin area has disrupted, i.e., either strengthened or weakened, the grid voltage of the apparatus, which corresponds to the absolute zero line.

It is easy to prove that it is actually the undamaged skin areas which causes the interference. For if one measures two abraded skin areas simultaneously, the absolute zero point does not move; the beam of light stays where it is.

THE POWER OF THE PYRAMID

The term "pyramid power" was introduced in the mid-seventies and alludes to the mesmeric properties of the pyramids at Giza in Egypt, specifically the great pyramid of Cheops. It has been demonstrated by a little group of researchers that a pyramid following a similar construction scheme (downsized) of the extraordinary buildings in Egypt will show comparable paranormal effects of its bigger counterpart.

Weird effects ascribed to pyramid power include the capacity to preserve food and embalm animals and plants buried inside. It has additionally been noticed that pyramids can protect metals from oxidation, entangle and redistribute life energy on the ground level and promote soil fertility.

Some Russian scientists recreated a pyramid on a large-scale using fibreglass in a series of experiments involving several pyramids, the largest of which is about 44m or 144ft high and weighing nearly 55 tons. The aim of the experiment was to look for the effects of pyramid power and methodically verify results in a number of key areas such as ecology, biology, chemistry and physics. The outcomes were incredible. After several trials, they confirmed that the pyramid was able to boost the immune system of animals and to promote growth in plants. A water solution, which had been instilled within the pyramid to alter its properties, was given to aggressive individuals and a marked reduction in their violent behaviour was noted. The same water was also given to drug addicts and a significant improvement in their abstinence effects was recorded.

An agricultural study was carried out on the effects of pyramid power on seeds and shoots. After placing test plants nearby the pyramid for varying periods of time, one to five days, it was found that when the seeds were sown, they could withstand drought much more effectively than the control counterpart kept away from the influence of the pyramid. An increase in crop yield was also noted ranging from twenty to a hundred percent depending on the type of seeds used. Subsequent experiments were conducted to analyse the effects on crude oil yield; a pyramidal structure was built on the location of an oil well in southern Russia. After a short amount of time, the overall production increased partially thanks to the reduction of the viscosity of the oil. It was also found that the chemical composition of the oil was molecularly altered, to promote the creation of beneficial polyphenol composites.

In literature there is a vast number of other studies on the effects of pyramid power, some of these showed that the levels of radiations decrease when a radioactive material is placed inside the pyramid, in a way similar to a reverse Faraday cage for EM radiation. It also seems that the amount of micro-seismic activity is reduced in the area of the pyramid. The level of toxic substances such as poison and heavy metals in food is reduced. The structure of crystalline substances such as salt and quartz growing are significantly altered after spending time inside the pyramid. Finally, a team of radar analysts witnessed a large column of ionised air swirling above the pyramid tip, which could be related to the turbulent ozone formation or to the Orgone energy flow.

From the numerous researches carried out, pyramid power has proven to be an exceptionally powerful phenomenon and can be employed to achieve a vast amount of seemingly miraculous yet scientific outcomes.

GIZA PYRAMIDS

The great pyramids of Giza have always fascinated multitudes of incredulous spectators in front of the immensity of these monuments of immense grandeur. Having maintained in excellent condition, their majesty symbolises the advanced complex of mathematical, engineering and spiritual knowledge of the ancient Egyptians. Entire villages were used for the construction and, in spite of the legends, the workers were not slaves but salaried workers.

What still strikes the collective imagination and does not find a precise explanation in modern Egyptology is the characteristic shape of the pyramids of Giza, with a square base with a *pyramidion* at the top, an element usually made of precious metallic material and polished like a mirror.

There is no official explanation for the pyramidal structure, the main hypotheses refer to the cone of light coming from the sun, to the primitive hill *Ben-Ben* on which the first human settlements in Egypt arose and to the reunification of the deceased pharaoh with the divine, in a resilient architectural thrust towards the sky.

Some scholars instead imagine there is a correlation between the triangular shape of the pyramid, seen from the side, and the symbol of femininity par excellence that is also found in the Christian tradition in the figure of the holy Grail, the V shape. In this case, however, the myth that the ancient Egyptians referred to was the cult of the *frog goddess*, a primeval divinity that generated the world, whose history is lost in the mists of time. Findings of small stone frogs in the vicinity of Egyptian burial sites may suggest the spread of the cult of the goddess even among the populations of the Nile.

It is not surprising that the frog symbolism was often found in ancient Egypt, as it was associated with gestation and childbirth. Back then, the adult frogs were numerous at dawn and their eggs abundant in the waters of the Nile. Since the Egyptians believed that life emerged from the primordial waters of *Nu* and that the Nile itself flowed from Nu, its waters would therefore have been indicative of new life, especially if it is considered that the river was essential to the growth of crops and to the sustenance of the Egyptian people. This connection is strengthened by the observation of the birth waters of the foetus that broke before the moment of parturition. Children, like frogs, emerged from water. The frog goddess, *Heket*, the divine midwife of ancient Egypt, the protector of new life, was often depicted as a frog or a woman with a frog's head and was frequently invoked to bring protection to the birth process or to defend the family unit and house guard. The amulets and beetles worn by women as protection at the time of birth often bore their image because it was believed that she brought relief to her mother and that her influence was manifested in infusing the first signs of life to a child still born, and in accelerating the last moments of childbirth. Many of the women who fulfilled such an important role were thought to be in service with her and in honour of their task of dispensing life they were called *Heket's servants,* thus the goddess became the protective deity of mortal midwives. Some midwives had the name or symbolism of the frog goddess engraved on the work tools they carried with them. Some sources describe this as a primitive deity as a daughter of *Ra* and say that she was born from her mouth next to the god of the air *Shu.*

A further explanation put forward by some authors is that these figures may have been sculpted with an apotropaic purpose, as an invocation to the gods. Placed next to the deceased or at the entrance of the houses, they served to protect the areas from malevolent entities. In support of this hypothesis, scientists affirm that some very similar figures are present in completely different parts of the world and especially in Ireland and Scotland, where they are called "witched stones". This hypothesis, like all the others, serves to explain many of the sepulchral statuettes often found in tombs but again does not cover all the aforementioned cases. These representations have always been found alongside the more advanced civilisations of history, such as Maya, Aztecs, Incas and the Egyptians, a fact that brings light on the deep connection between the frog goddess and the Orgone energy, that was well understood and used by all these civilisations. Furthermore, the geometric ratios between the base and the height of the pyramids hide an essential project. All the previous hypotheses bring back to the imitation of nature, excluding the possibility that instead the pyramidal shape was chosen because of its peculiarity: the ability to capture environmental energy.

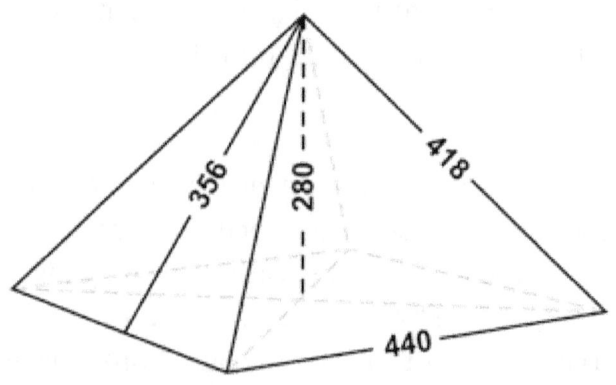

measurements of *Cheope's* pyramid

In the late thirties of the last century, the brilliant Austrian scientist disciple of Freud, dr. Wilhelm Reich, connected the dots and explained the pyramid power in terms of Orgone energy modification. This energy, accumulating locally, is able to polarise the water molecules and, consequently, modify the weather; not only, by conveying the Orgone in a closed room it is possible to cure diseases, increase the overall well-being of the individual and, according to historical accounts, to prolong human existence to levels that medicine has not yet been able to reach in any way.

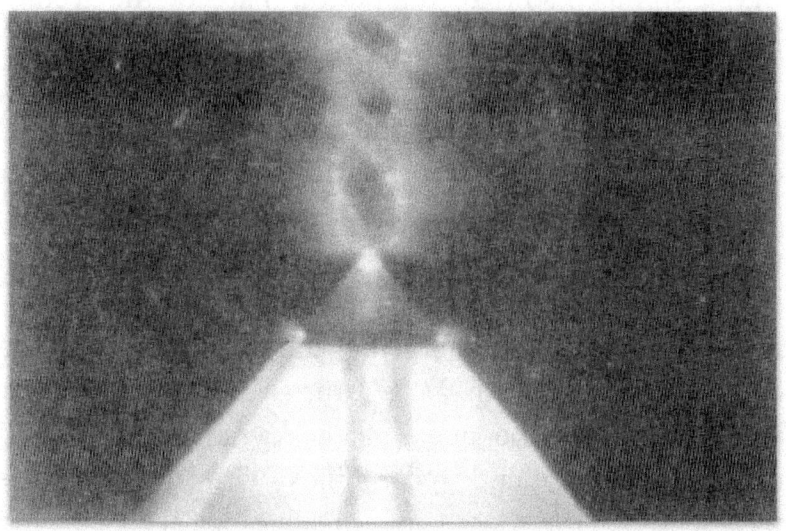

kirlian photograph of the energy field produced by a pyramid excited by a Tesla coil

Recent studies carried out following Reich's work also by myself prove that it is largely the pyramidal form that guarantees an increased flow of Orgone energy. Flowing through the tip into the heart of the pyramid and then exiting from the lateral surface, the Orgone is collected by the stratification of organic materials (rocks and minerals) and inorganic (metals) and strengthened, as can also be verified by comparing the ionisation of the air at the base of the pyramid and at the tip.

The fact is that it was precisely the pyramids and their constant release of purified orgonic flow to guarantee prosperity, wealth and health to the people of Egypt, so much so that in a period that saw the average life for a man to be around 35-40 years, in Egypt there were numerous cases of late deaths. For example, Ramses II was mummified when he was exactly 90 years old. It is not to be excluded, but it would require further studies, that the very mummification and burial within these structures was due to the profound knowledge of the vital energy of the Egyptians: perhaps they understood the all-embracing nature of the orgone energy and saw death as a return to the energy flow itself that from the apex of the pyramid is released towards the sky and recombines with the Universe transcending the death of the biological body and leaving a strong imprint in the local energy field.

It is also interesting to note that, regardless of the materials used, any pyramidal shape can convey energy (with low efficiency), so much so that even the Romans built numerous pyramids in Rome during the brief period of Egyptomania and it was precisely that period that was one of the most flourishing for the eternal city. Unfortunately, all of this was of short duration since, unlike the Egyptians, the Romans never really understood the correlation between the pyramids, the cosmic energy and the flourishing of the city.

Piramide Cestia in Rome

Moreover, the phenomenon of orgonic energy convergence is accentuated if we find ourselves in the presence of water, so much so that Reich himself repeatedly demonstrated that nearby streams and rivers the energetic potential was clearly higher than the average: this is due to the water cycle and to its filtering through rocks that act as a natural accumulator being composed of layers of organic material (limestone and vegetation) and inorganic (ferrous minerals) enriched with energy.

In fact, both Giza and Rome are located near a large river, respectively the Nile and the Tiber.

Such occurrence lead to study the relationship between pyramids and surrounding areas; the Nile river exerts a considerable influence on the surrounding environment, at the time of the powerful pharaohs the floods were regular and often waters came almost to the base of the hill on which the pyramids lie, enhancing the local energy flow: this is the reason why the Egyptian land has been among the most fertile on the planet for centuries.

All of this was abruptly interrupted precisely with the fall of the Egyptian empire when the monuments looted, deprived of the metal components and of the white marble covering, started to lose power, with consequent drying up of the territory, advancement of the desert and migration of the population towards Mediterranean shores. Meanwhile, the great discoveries of a millennial were dispersed among the sands of time and forgotten for centuries, recently rediscovered yet opposed by the academic and political world, so much that Reich himself was imprisoned and his researches burned.

Orgone is now seen as a pseudo-science, despite the copious evidence and experimental confirmations of numerous independent academics, and research is often sabotaged or covered up. Nonetheless, not everyone forgets the teachings of the Ancient Masters hoping one day to have a society in which the well-being of the majority prevails on the greedy desire for power of a few.

A BRIEF OVERVIEW OF ORGONOMIC DEVICES

THE ORGONOSCOPE

Among the first devices developed by Reich there is the *orgonoscope,* an instrument able to make Orgone visible to the naked eye. The orgonoscope is a tube made with different materials to trap Orgone inside and with a pair of lenses acting like an accumulation chamber. Looking at the night sky from the lenses, a disk of brighter blue will appear together with a flickering movement due to orgonic energy pulsations.

Diagram: Orgonoscope

C: cellulose disk, outside surface dull
W.M.: wire mesh, on both sides of disk
M: metal cylinder, about 4″ long, 2″ wide
L: biconvex lens, about 5x, focused on disk
T: telescopic tube, 1 to 2 feet long, about 2″ wide
E.P.: eye-piece, 5-10x, for additional magnification

"Keeping both eyes open and peering through the tube with one eye, we see a dark-blue night sky within which is a disk of brighter blue. Within the disk itself we perceive, first of all, a flickering movement, then, unmistakably, delicate dots and streaks of light appearing and disappearing. The phenomenon becomes less distinct in the immediate vicinity of the Moon; the darker the atmosphere in the background, the clearer the phenomenon". (Reich, *The Cancer Biopathy*)

Reich performed most of his experiments in either Oslo (Norway) or Maine (USA). Both of these locations are geographically very close to the north and much more susceptible to the interaction of cosmic rays with Orgone. In further orgonoscope developments Reich added a cellulose disk with calcium sulfide or zinc sulfide, two phosphorescent substances acting as scintillators. The disk was added to perform more precise measurements of colliding bions and caused small flashes of light when detecting an incoming particle. More precise measurements were performed placing the disk under a microscope, especially if the whole experimental apparatus was in the near vicinity of an ORAC. Cosmic rays and photons are too energetic to interact with the disk and can't be held cause of possible false detections.

THE ORGONE METER

In the meantime, Reich had been tinkering with a new apparatus able to quantify Orgone presence in the environment inasmuch as probe its charge.

In Reich's own words: "The different pole of the secondary coil of an induction apparatus (an old diathermy apparatus, for instance) is connected by ordinary electric appliance wire to an iron plate, 2 feet long and 1 foot wide. The iron plate is insulated on the underside with wood. A similar metal plate is then mounted above and parallel to the first at a distance of 6 to 12 inches and in such a way that it can slide up and down. The top side of the upper metal plate is insulated with a piece of plastic celotex, or like material the same size as the plate and 1/2 inch thick. Electric wire connects the two iron plates to a simple cylindrical bulb of about 40 Watts set between them". (Reich, *The Cancer Biopathy*)

The Orgone meter acts as a direct measuring device using electromagnetic properties of charged particles passing in an electromagnetic field causing a speed vector variation in direction (but not in modulus). It is the same principle of mass spectrometry where the time of flight is employed to determine the mass of an unknown particle.

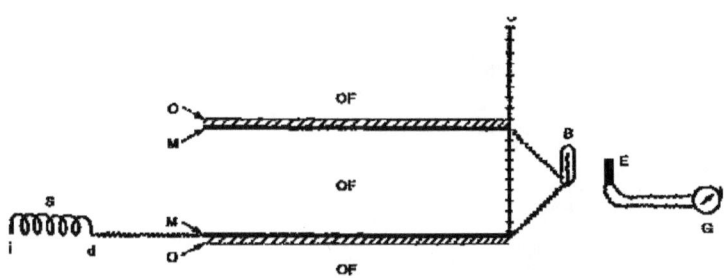

A working scheme of the Orgone meter

THE CLOUDBUSTER

A very innovative device Reich invented is the cloudbuster, featuring several hollow long metallic pipes all lined up and pointing in the same direction. They converge into a very long tube submerged into a river nearby. Since water is a natural Orgone absorber, the metallic pipes act as focusing elements to rebalance the Orgone presence in the sky. A direct consequence is the increase or prevention of rain and storms.

His revolutionary theories attracted considerable media attention and in 1953 blueberry farmers in Maine offered to pay dr. Reich if he could build a device able to end a severe drought that threatened their crops. Reich set up his cloudbuster (whose remains are still visible) and operated it for just over an hour; the next morning it started raining and the crops were saved.

During the last decade or two, Reich's discoveries on the presence of a persistent atmospheric energy continuum, which causes most climatic phenomena, have been confirmed and reinvigorated. Reich's weather modification device, the cloudbuster, has been used to stop numerous storms, heavy winds, drought periods and once in a while, to stimulate rain into deserts. Such device surely proved to have a ground-breaking impact, and this has been demonstrated in several trials. When appropriately used, it can trigger enormous separated tempests, or even modify the whole storm front, provoking huge rainfalls or mitigating strong winds. Wrong use or construction errors may cause an exacerbation of extreme conditions leading to floods, wildfires and hailstorms.

Such device has the capacity of affecting the Earth jet stream and controlling enormous tempests: even small hurricanes can be deviated given a properly set cloudbuster. The main prospected use is the reestablishment of common cycles of precipitation in regions afflicted by long dry season or desert.

However, as experienced by Reich himself and a lot of other important figures in history, groundbreaking discoveries are never really understood at first. Governments, strong powers and economic oligarchies oppose innovations that can mine the status quo, mainly by leveraging the scientific community against using its natural resistance to change. In fact, by adding several "friendly lies", consistent works can be manipulated to the point of becoming totally deviated and fake.

Examples of such behaviour are bringing in the UFO, the flat Earth, government conspiracies and similar unlinked theories just to damage the author with a constant *reduction ad absurdum*.

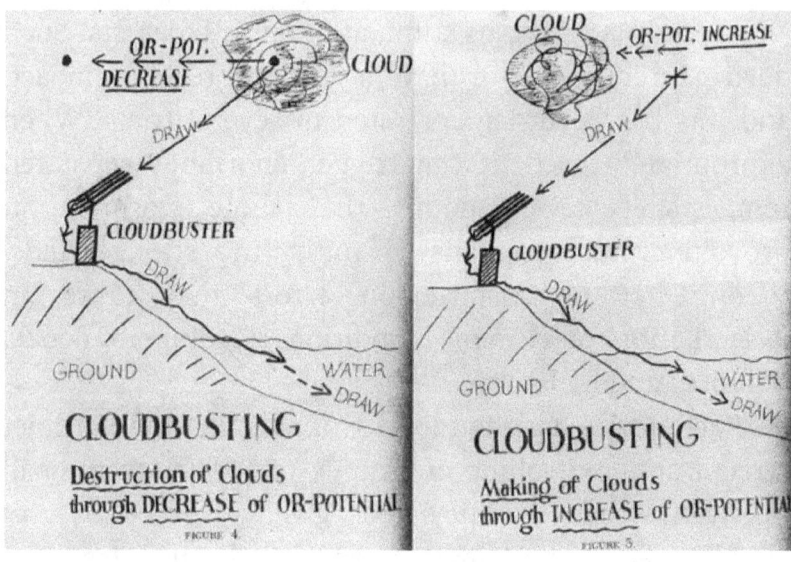

It should be noted that the cloudbuster is an instrument whose main aim is to reestablish a state of self-regulated atmospheric condition. It is completely different from climate alteration techniques, such as cloudseeding, where the environment is forced to do something it regularly would not do.

THE ROTORGON

The Rotorgon (a name inspired by the rotor powered by Orgone) is a basic device created by Carlo Splendore to identify the presence of an existing orgonic energy field, both radiated by the human body, present in the environment, or coming from astronomical emitters and passing through the Earth. The main part of the instrument is an Orgone collector, a fundamental element as it can give insights about specific properties of the Orgone under analysis. In the Rotorgon, life energy interacts with the inner reaction chamber translating electrostatic energy into actual motion. The rotor will start to spin depending on the amount of Orgone detected; looking at the spinning direction, the instrument can identify the carrier charge and distinguish between POR and DOR. Some researchers also think it could be hacked in order to create a real motor using Orgone to produce kinetic labour, thus acting as a renewable energy extractor with considerable ecological benefits.

Such thesis is still questionable due to the very limited amount of power harvested by these types of devices but research is going on to improve the overall conversion efficiency.

ORGONE THERAPEUTICAL BLANKETS

Orgone blankets accumulate orgonic energy naturally without the use of batteries or hazardous electronic devices. They stimulate the parasympathetic system response by relaxing the body and its muscles while accumulating life energy inside enriching the body tissues with bions. Effects can range from a simple stress relieve to a faster healing and immune system boosting.

Those blankets are made of several layers of wood and metallic shavings, each layer carefully encapsulated to concentrate the environmental Orgone charge as well as block the energetic depauperating process of the body. The number of layers is strictly connected to the effects wanted and the type of blanket, whether a cold winter or fresh autumn one.

THE BOVIS BIOMETER

Despite not directly related to Orgone and dr. Reich, it is worth talking about the Biometer and the Bovis scale of radiant energy.

Invented by the French researcher Alfred Bovis with the help of Andrè Simoneton, a famous engineer of the time, the Biometer is a new method of calculating the energy emitted by a man, an object or by the environment. It consists of a chart with a suspended pendulum oriented to the Sun (or the Moon at night). Its position is a good indicator of the radiant energy in the immediate surroundings. Vital vibrations are measured according to a precise scale, the Bovis scale, in Ångstrom (one ten-billionth of a metre or 0.1 nanometre) since it represents the energy wavelength.

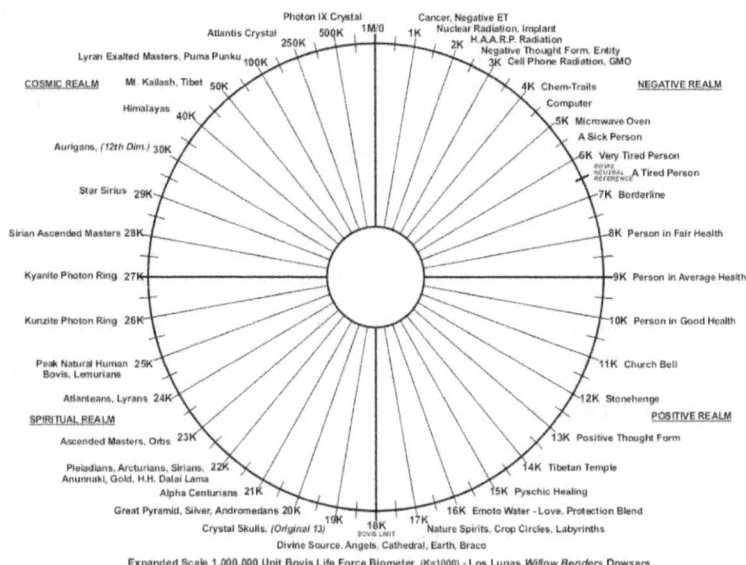

BOVIS LIFE FORCE BIOENERGY UNITS DOWSING CHART

Expanded Scale 1,000,000 Unit Bovis Life Force Biometer, (K=1000) - Los Lunas *Willow Benders* Dowsers
May be copied for personal or chapter use. www.loslunasdowsers.org Revised 4/01/14, Gary R. Plapp

ORGONE AND MATTER

The choice of materials for construction of anything is fundamental, just think of a house build with sand instead of concrete subject to extreme weather conditions. The same applies to orgonomic devices, since they are actively interfering with the human body, quality and efficiency must be priority guidelines.

The creation of any device interacting with Orgone energy is essentially the superposition of natural, dielectric materials substituting with metal materials, with a metal layer nearer to the organic body to reflect and direct the energy stream towards the person inside. The path of bions is consistently from the outside natural layer to the inside which is secured with a metal-based material. Orgone collectors are made in different shapes and sizes, out of various materials, and in this manner serve various uses depending on the effect pursued.

Materials used to develop an Orgone collector are very important as they will contribute to the nature of the orgonic charge. Reich employed wax or fibreglass for his natural layers since he found these materials to have a positive charge, yet they additionally did not show a shielding effect which can diminish the overall energy flux. From tests polyesters, vinyl, polyurethane, ABS, neoprene, nylon and styrofoam were observed not to be great materials to use. They cause excessive shielding and scattering of bions which results in a very reduced efficiency of particles polarisation. Likewise, Reich found that it is imperative to just employ iron-based metals, for example, carbon steel or pure iron sheets.

The human haemoglobin atom, responsible for blood colour, is made out of Carbon and Iron, and carbon steel (iron) is evidently perfect for charging human blood inasmuch as it can interact with any organic element. Different metals, for example, aluminium or copper, can destructively interfere with Orgone bions carriers charge and ought not be taken into account for construction.

For an Orgone accumulator, the quantity of active layers of natural and metal material makes for the quality of the total energy collected. The bions pressure within the chamber can be emotionally felt and furthermore can be estimated by temperature changes and with the help of a simple foil electroscope. Yet layering isn't the only guiding rule, as for just 3 layers can be as effective as 10 layers if materials are carefully prepared, placed and chosen.

CRYSTALS

Crystals are a fundamental part of Orgone devices and show several macroscopical properties such electrical conductivity, magnetic properties, and the lattice structure are very susceptible to electromagnetic interference. Fe and Al have an unevenness of positive charges in the atomic structure in the external shells, making them have unpaired electrons. Better said, the presence of electrons in the outer orbitals that has not been filled subshell causes Fe and Al to have unpaired negative charges. The movement of unpaired electrons in the lattice structure also leads to both iron and aluminium being good conductors. Electrons in this case can be considered like a gas permeating the whole atomic structure (the Fermi gas model of metals). Electrical conductivity is a significant property of a material linked to the internal dipole magnetic field. At the point when an electric field is given in a dielectric, polarisation will manifest on the dielectric surface.

However, according to the undergoing examinations, the absorption of electromagnetic waves in iron-enriched Orgonite is superior to aluminium-enriched Orgonite. This is grounded on the fact that iron shows strong magnetic properties, being ferromagnetic, while aluminium is paramagnetic, or weakly magnetic. Orgonite phenomenon starts with the induction of friction-based electricity in the crystal. This characteristic is known as the piezoelectric effect of quartz. The piezoelectric effect will create an electric field due to the shape change and consequent relocation of electrons in the molecular structure. The higher the shape deformation is, the stronger the resulting electric field is. The resulting electric field of crystal polarises the absorbed electromagnetic radiation and transforms it into heat, thanks to another phenomenon called electrostriction. Then, energy is released through the tip of the crystal adjacent to the positive terminal with a positive charge, thus showing life-enhancing capacities. In parallel, the magnetostrictive (shape deformation due to magnetic fields) effect of metals compresses the quartz crystal and consequently enhances the piezoelectric effect in what looks like a perpetual resonant cycle.

Based on this study, it was obvious that Orgonite has the potential to moderate electromagnetic radiation in environment, so as to safeguard the effects of radiation on human health. Orgonite can be made directly on a household scale because the materials used are cheap, easy to find, and easy to process. Orgonite aesthetics can be improved with attractive design and the addition of objects which also serve as absorbers for the full spectrum of electromagnetic radiation generated from electrical appliances such as cellphones or computers.

QUARTZ

Despite being a crystal, it deserves a section on its own for its marvellous properties.

Quartz is a very common material on earth and exists in many different forms, amethyst, ametrine, aventurine, citrine quartz, blue quartz, tiger's eye, agate, jasper, etc. It is made by silicon and oxygen atoms ($SiO4$) in a specific geometrical pattern. The mineral itself can be divided into two main categories, crystalline quartz created by the addition of other silicon oxide molecules layer by layer:

- Cryptocrystalline quartz showing infiltrations of other materials recognisable by the fibrous and grainy structure;
- Crystalline quartz shows the best purity and transparency, especially in the form of single or double terminated, which means that one or both terminations are naturally pointy and haven't been artificially carved.

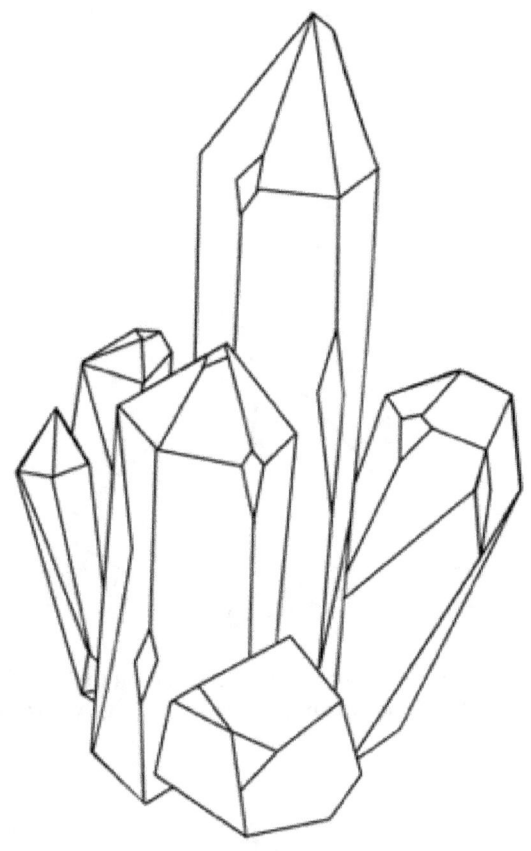

Before usage the crystal should be cleaned under cold water and left for a day in full sunlight.

Good positioning of the stone is very important to funnel Orgone from the tip of the device to the quartz crystal. The shape and geometrical constraints give a hint on the direction of the Orgone vortex. The crystal vertex should point upwards aligned with the pyramid axis.

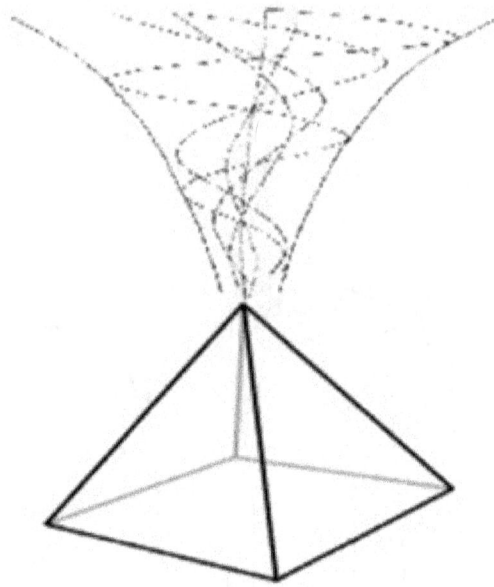

The image shows the Orgone vortex created on the tip of the pyramid
made of several turbulent waves of energy converging.

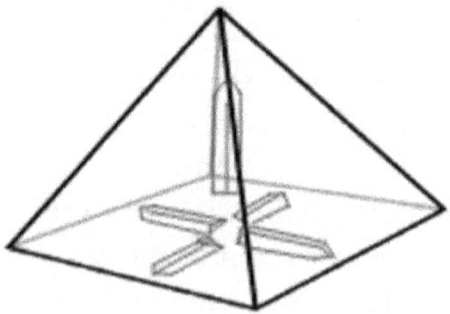

If using multiple crystals, they should be placed following the cardinal
axis and the four directions (east, west, north, south).

RESIN

The common binding agent in Orgonite is the resin, a type of plastic made of long chains of molecules repeated over and over, the polymers. There are mainly two types of resins, the polyester and the epoxy.

Polyester resin, the main component of fibreglass, is produced by the reaction of resin with a catalyst that triggers the polymerisation process. The catalyst percentage is usually from 1% to 3% in volume. The best working temperature for this type of resin is 20°C or 68°F.

Epoxy resins are used in the assembling of wood, plastics, paints, coatings, glass and sealers, floors and different items and materials that are employed in construction and artistic applications. They are a kind of thermoplastics created by the mixture of at least two components, often named with a capital letter (part "A" and part "B"). Most glues sold as "auxiliary" or "designing" cements are epoxies. These superior types of glues are perfect to make a plastic cover for decks, dividers, rooftops and other structural applications. Epoxies can work with wood, metal, glass, stone, rocks and a few plastics, and many more consumer materials. Epoxy resin additionally helps to give a strong, polished outer coating, just as sealers for floors or englobing agent for flowers. Mixing the two components must follow precise weight ratio rules different from product to product: for an effective use it is required to look at the instructions given from the resin producer.

The resin will solidify in 1h to 4h plus a day of consolidation. It may overheat up to 50°C or 122°F in the process. Ventilation of working areas is required during the whole process. It is very recommended to make small amounts of resin as the leftovers can't be reused.

Polyester resin shall not be used as it isn't as clear as epoxy, is prone to cracking and is extremely toxic. A common belief related to polyester resin in Orgonite is that the plastic shrinks during the curing process, permanently pressing the quartz crystal inside causing a well-known piezoelectric effect inside the crystal, so that its termination points become electrically active. This is totally irrelevant as a piezoelectric effect works by constantly squeezing and relaxing the material; instead, the effect responsible for Orgonite inner working is the magnetostrictive effect caused by electromagnetic waves and Orgone itself on the crystal, wrapped in a electromagnetically resonating copper spiral.

The perfect mix of resin and metal is of 50/50.

METAL

First of all, it is worth saying metal powder can't be used because it causes several problems when incapsulated in the resin. The same can be said for greasy and dirty metal pieces, that won't perfectly blend into the epoxy and may cause bubbles or cracks. Metal parts in general should be properly cleaned with alcohol in order to remove possible contaminants. Be careful of dubious metal scrapings, they might contain plastic parts.

Special attention should be placed to copper spirals too. A spiral is a concentric resonating element that, paired with the quartz crystal and the Orgonite, can cause two main effects:

- Picks electromagnetic waves and leveraging the piezoelectric and magnetostricic effects of quartz, removes such frequencies from the surrounding environment;

- Cleans out spurious waves and noise that are too weak to induce any effect in the crystal;
- Tunnels electromagnetic pollution into the device to further increase inner Orgone charge and stimulate the DOR-POR conversion.

It is possible to calculate the frequency the metal will be able to pick and emit. For instance, a 1cm or 0.39" piece of metal can resonate at frequencies starting from 7.5GHz and above whereas a 3cm or 1.18" one with 2.5GHz and above.

The empirical formula is

Length (meters) = 300 / frequency (MHz)

You can then determine the required minimum dimension of the copper wire according to the frequency to be removed.

Frequency Band Name	Frequency Range	Wavelength (Meters)	Application
Extremely Low Frequency (ELF)	3-30 Hz	10,000-100,000 km	Underwater Communication
Super Low Frequency (SLF)	30-300 Hz	1,000-10,000 km	AC Power (though not a transmitted wave)
Ultra Low Frequency (ULF)	300-3000 Hz	100-1,000 km	
Very Low Frequency (VLF)	3-30 kHz	10-100 km	Navigational Beacons
Low Frequency (LF)	30-300 kHz	1-10 km	AM Radio
Medium Frequency (MF)	300-3000 kHz	100-1,000 m	Aviation and AM Radio
High Frequency (HF)	3-30 MHz	10-100 m	Shortwave Radio
Very High Frequency (VHF)	30-300 MHz	1-10 m	FM Radio
Ultra High Frequency (UHF)	300-3000 MHz	10-100 cm	Television, Mobile Phones, GPS
Super High Frequency (SHF)	3-30 GHz	1-10 cm	Satellite Links, Wireless Communication
Extremely High Frequency (EHF)	30-300 GHz	1-10 mm	Astronomy, Remote Sensing
Visible Spectrum	400-790 THz ($4*10^{14}$-$7.9*10^{14}$)	380-750 nm (nanometers)	Human Eye

WAX

Some attempts have been made to use wax or natural resins to replace resin. The main drawback is that the resulting device will be very fragile and will never really settle down, so abrupt impacts or temperature changes may damage it.

Experimenting with wax is pretty simple, just replace the resin with molten beeswax. Resulting pyramid will be very sensible to Orgone oscillations and may even explode when reacting to strong local orgonic energetic fields. As for natural resins, they will not be discussed further in this book.

<u>MOULD</u>

It is advisable to buy silicon moulds as their flexibility makes de-moulding very easy. They can be bought on most online and offline shops in all sizes and dimensions. Stainless steel moulds are more difficult to use and need to use an appropriate detaching agent sprayed on all internal surfaces before placing anything inside. Be careful about plastics that cannot hold high temperatures because polymerisation is an exothermic reaction creating temperatures up to 50°C. Lastly, glass should be avoided as mould material because it is fragile and tends to bond with resin.

Silicon moulds for pyramidal Orgonite

MAKE YOUR OWN ORGONITE

SAFETY NOTICE

The building process includes the usage of dangerous chemicals, please use gloves and protection glasses and work in a well ventilated area. Liquid resin can be polished with acetone while hardened one is nearly impossible to remove, make sure the working space is covered to prevent spills and keep children and domestic animals away.

PREPARATION

First step for both an Orgone accumulator – ORAC replica - and an Orgonite purifier is to plan the size and the shape as well as to think about the location. This is especially true if you want to sit inside the accumulator so that the whole structure needs enough space for a person to comfortably sit inside. For the purifier, the shape is limited to using the schemes of either one of the Giza pyramids or to use a perfect geometrical form. Both have advantages and disadvantages, namely the lack of ready-made silicon shapes off the shelf for Giza pyramids. A perfectly equilateral shape is preferred when dealing with purifiers used inside homes as they are better at dealing with Orgone coming from the sides, whereas a Giza-shaped pyramid is better at intercepting and trapping Orgone coming from the sky and outer environment.

Next step is to gather materials, preferably directly from nature when it comes to rocks and crystals. Metal and steel wool can be taken from old structures or savaging non-electrical equipment. An example is metal shavings recycled from old mechanical parts.

MOULDING

1. Put the mould upside down on a solid surface, it can be a glass or any other container that can hold it still. Place some tape on the contact points to block possible movements of the pyramid. Apply a releasing agent on the inner side of the mould to prevent resin from sticking to surfaces.

2. In the meantime, prepare a small amount of epoxy mixing the two components according to the ratio specified on the instructions from the vendor, don't mix too vigorously to avoid incorporating air bubbles. Use a container suitable for high temperatures as catalysis of resins is an exothermic

reaction reaching temperatures up to 50°C / 122°F.

3. Now fill the top of the mould with steel metal shavings.

4. Add a single or double terminated quartz crystal wrapped in a copper wire spiral on top of the shavings with the termination facing the pyramid tip and cover with epoxy. Then, on the sides, metallic disks, copper spirals or simply other metal shavings can be placed. Add epoxy.

5. The bottom of the pyramid (which in this setting is at the top of the mould) should be covered with iron and copper parts mixed with crushed quartz crystals, almost to the point of being a powder.

6. Leave the mould in a dry and warm place avoiding direct sun exposure for an entire day.

<u>POLISHING</u>

Gently take the pyramid out of the mould, be careful not to break the tip. Then rub the surface with soap and water to remove any trace of the releasing spray. It is preferable to avoid degreasers and simply use a dry cloth or some soapy water instead.

Dry cloth cleaning should go from the tip to the base in a circular motion and different cloths must be used for each device not to contaminate one another with electrostatic energy.

USAGE AND MAINTENANCE

Orgonite pyramids can be placed everywhere but in order to maximise their potential a few rules must be followed.

○ Inside a typical apartment, you may want to put it near sources of EM pollution (TV, microwave oven, WiFi router, etc) and, at the same time, close to people. Likewise, they can be stored near water pipes benefiting from the natural flow of Orgone through water despite shielded from the tube;

○ Near windows so that indirect sun exposure will always fill Orgone levels within the device. Direct and prolonged sun exposure may ruin the resin;

○ Outdoor placing is also possible but the device must be kept away from environmental agents like rain and sunlight.

A common misuse is interring Orgonite nearby cell towers and energy power-plants hoping to counter their negative influx: such approach is totally flawed as the enormous amount of emissions coming from those elements cannot be countered on a whole area just using a simple Orgonite. It is best to live away from such structures or use an Orgonite pendant to mitigate constant EMF exposure. Burying resin-based objects is also environmentally obnoxious because plastic will slowly deteriorate and pollute the surrounding area.

Always remember to recycle your Orgonite as plastic or to donate to someone in need, not only you will help another person to benefit from the power of Orgone, but also avoid wasting precious materials.

Since Orgone collectors distill environmental vital energy, the nature of the encircling air is critical for a positive outcome. Orgone aggregators work best on dry, windy and bright days when the energetic charge in the environment is positive and growing. The timespan from early morning to early afternoon appears to be most suitable because stale air from human activities is still relatively low. On rainy or cloudy days, the Orgone energy flux is held in the upper part of the atmosphere, so that its presence at the surface level will be much less. Orgone energy is likewise less present during the evening when the Sun's direct radiation is lighter and the atmosphere polluted by anthropic influence.

It is important to notice that the aforementioned devices can be used under almost any condition, what really changes is the overall efficiency. A remarkable exception comes from those days with extraordinary fogginess followed by warm dry weather: wildlife and plants appear to be drowsy, inanimate and slow. Reich explained similar events based on the presence of a stagnant Orgone bubble that has turned into DOR. Desert areas seem to be highly affected. During such days it is better not to use any device and to avoid excessive outdoor activities or meditation.

Both Orgone accumulators and Orgonite purifiers are subject to an effect called efficiency degradation, which is the loss of effects over time due to the consumption of the inner layers and the accumulation of negative ions on the outer surfaces. Such degradation can be detected simply looking at the surface for any blackening/whitening (or, in general, colour change) or dust accumulation.

The first cleaning procedure takes advantage of the deionising effects of still water. Simply washing the device under flowing water or wiping it with a water-soaked cloth is sufficient to re-establish a correct carrier charge on the surfaces. On average, any device interacting with Orgone should be cleaned every month.

A second common maintenance practice, rather easy to perform but with enormous benefits to slow down layer corrosion, is to leave the device under the light of a full moon, when the sky is clean and very dark. Artificial lighting can interfere. The process is strictly connected to the photoelectric effect, although happening with the much less energetic scattered photons coming from the Moon.

A very dangerous form of decay happening to accumulators is the DOR pollution. As Reich himself noticed during the Oranur experiment, sometimes a polluted environment can cause the spontaneous and irreversible transition of the once-beneficial Orgone accumulator into a deadly Orgone trap. Such a device would no longer extract beneficial energy from the environment yet only attract negatively charged orgonic energy, creating harmful effects within the chamber for organic beings. An example is the famous FDA test on an Orgone box where they used an X-ray machine to detect a possible energy flow. Nothing was detected and the accumulator was declared to be worthless yet what really happened was well beyond their reach: the ionising radiation in the immediate contingency blocked the machine from working and reversed the orgonic energy flow, causing an undetectable DOR contamination.

Similar effects can happen in cities where a smaller degree of DOR pollution may interfere with the devices causing a general weakening of the effects yet not reversing the functionality to a harmful level. This is due to the several forms of pollution cities are soaked in, a topic discussed in the appendix of this book. In the *Orgone Energy Bulletin* (Oct. 1952) Reich described DOR clouds as deadly accumulations of negative life energy in the environment resulting in a negative orgonic potential field with strong effects on the surrounding area. First, it impairs vegetal and animal life causing desertification and death of small animals. Secondly, people will experience negative feelings when passing by, the most sensitive accusing flu-like symptoms with no apparent reason.

TESTS

Properly testing an Orgonite is a fundamental step to ensure the building process and the preparation have been successfully performed. It is also a vital indicator of the sometimes inscrutable inner workings of Orgonites and Orgone accumulators, especially if acquired online or from an unknown source.

Direct Tests

In this way we test the presence of Orgone around the device and denote the overall efficiency. This kind of analysis often uses expensive equipment so for a DIY builder the main route is to try indirect testing. Moreover, being weakly interactive, direct measurements of Orgone energy are very likely to fail unless repeated several times using extremely precise and accurate instruments.

Indirect Tests

These types of analysis are mainly looking at the effects of Orgonite on the surroundings, be it water or living beings. A very simple way to see if the orgonite works or not is to use its effects on water solidification.

Simply putting an orgonite on top of a glass of water in the freezer can show a vortex pattern appearing inside the ice.

Another test often performed with both Orgonite and Orgone accumulator chambers is to place 5 to 10 lentils on some humid cotton inside a petri dish and to place one inside the accumulator or near the Orgonite while the other in another room keeping the same environmental conditions (same temperature, light exposure, humidity, etc.). Then note the difference after a few days.

Air polarisation and EMF blockage are also good indicators of an effective working of the device. Both ion and EMF detectors are sold quite inexpensively online. For the test simply put the ion detector nearby the tip of the pyramid and take a second measurement a few steps away. The ionisation at the tip should be much greater, a sign that the Orgone flux is being gathered at the top of the device. In order to see actual EMF just place the pyramid between an electromagnetic radiation source (Wi-Fi, mobile phone or cordless, for instance) and the detector. Then take the pyramid away. The measurements should highly differ, a sign that the Orgonite is actually acting as a proactive shield, both blocking radiation and converting it into electrostatic energy to perform the inner DOR-POR reaction previously discussed.

A third experiment very easy to interpret is the ice test: place an Orgonite on top of a glass of water and put everything in the freezer for 12h to 24h without touching or disturbing it. Ice inside the glass will show an inner vortex shape due to the Orgone energy flow interfering with the water solidification process. It is worth mentioning dr. Masaru Emoto research on different water structures formed during crystallisation from liquid to solid (ice). Ice crystals create diverse aggregates according to external influences, from shapeless clusters of water molecules to highly-structured crystal lattice with hexagonal arrangement.

Finally, a Biometer and relative Bovis scale are excellent indicators and, if properly used, can give a qualitative (and quantitative in lab conditions) measurement of the devices efficiency and emitted radiant energy.

THE (IN)VISIBLE
ENEMIES

Nowadays human health is endangered by many new threats, most of them man-made. The immune system needs to keep up with the microscopic assault of many pollutants like micro-plastics, Electromagnetic Fields (EMF) and heavy metals.

Humans have evolved and developed in a natural environment where health hazards were represented by plant toxins, wounds and germs. Things like strong Electromagnetic Pulses (EMP) and small plastic particles, as well as drugs and artificial substances with dubious effects on long-term are a completely new thing. Studies ongoing show how dangerous these contaminants can be in the long-term, since the body itself is totally unable to tackle with their presence and the main form or reaction is tumour, i.e. a malignant alteration in the cell structure causing its uncontrollable replication.

ELECTROMAGNETIC POLLUTION

With the impressive increase of electronic devices in proximity of cities down to being in direct contact with our body (hands up who keeps a smartphone in the trousers), electromagnetic pollution has become a relevant issue. EMF have shown to interfere with cell reproduction and, if powerful, can afflict irreversible damage to organic tissues. But let's take a step back and see what really makes an electromagnetic wave.

In physics energy assumes many forms, from heat to movement to light. The latter, in particular, is actually made of infinitesimal particles named photons that travel at the speed of light even through vacuum and carry a certain amount of energy according to their velocity. Now in quantum mechanics every elemental particle can be seen as a wave, thus photons and electromagnetic energy can also be represented by a wave whose frequency is intrinsically linked to the photon energy and speed. Higher frequency means higher energy carried.

The electromagnetic spectrum depicted below ranges from waves as big as buildings (long radio waves) to gamma rays coming from outer space, the size of an atom nucleus.

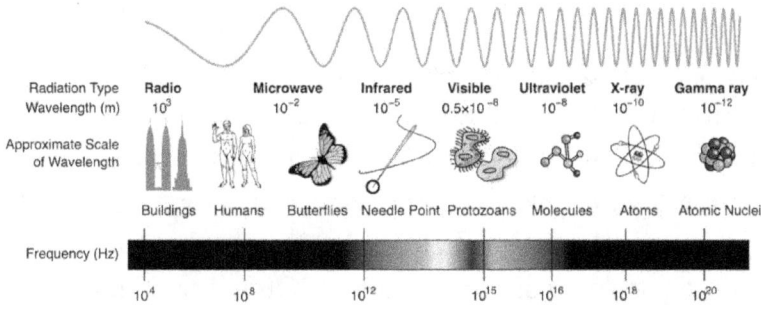

Radiation Type	Radio	Microwave	Infrared	Visible	Ultraviolet	X-ray	Gamma ray
Wavelength (m)	10^3	10^{-2}	10^{-5}	0.5×10^{-8}	10^{-8}	10^{-10}	10^{-12}
Approximate Scale of Wavelength	Buildings	Humans	Butterflies	Needle Point	Protozoans	Molecules	Atoms	Atomic Nuclei

Frequency (Hz)

10^4 10^8 10^{12} 10^{15} 10^{16} 10^{18} 10^{20}

Now looking at the spectrum of EM waves it is very noticeable how even Wi-Fi and microwave oven emissions carry a considerable amount of energy. If microwave and infrared just cause overheating of tissues, UV, X-Rays and Gamma Rays are extremely dangerous for biological cells causing DNA mutations, cellular decay and cancer. Yet, even overheating of cells, that is thermal contamination, can also cause problems to the organism.

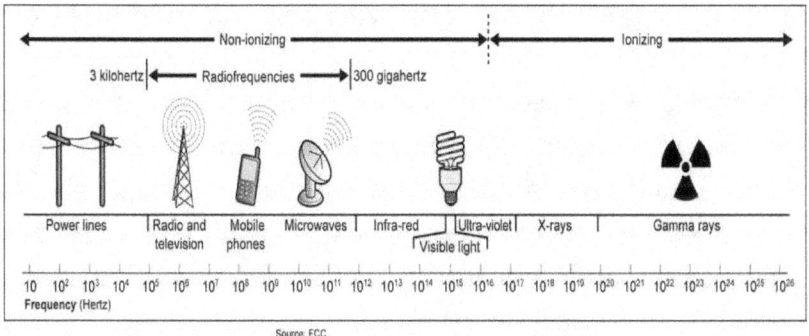

Source: FCC.

Nonetheless, interaction with organic bodies is detrimental in terms of both energy dissipated and continuous exposure as the sources are always present and active 24/7 for 365 days a year. Not only EMF can disrupt organic life, but they can also interfere with the local Orgone energy fields. A strong electromagnetic field can turn Positive Orgone (POR) into Negative or Deadly Orgone (DOR) exciting the bions: a very tangible result is the wasteland usually seen around high tension power-plants and telecom infrastructure so that even plants seem to grow less in the neighbourhood of such places.

A DEADLY INNOVATION

When it comes to technology there seems to be no stop button or damping system to block engineers from putting organic lives at risk with new untested systems. This is the case of 5G.

In 2018 a research study founded by the United States government concluded that there is clear proof that radiation coming from mobile phones shows cytotoxic effects, explicitly, causing heart tissue disease in rodents and cancer. In the study, involving many rodents and mice (whose natural similitudes to people make them valuable markers of human health dangers) to dosages of radiation comparable to a normal mobile user lifetime exposure. It's interesting to notice in most countries mobile phones were introduced without any proper safety testing. Nonetheless, the mobile phone is a device we mostly keep closed to our bodies for a prolonged time, containing RF trans-receiver circuitry. A very famous image shows how 15 minutes of calling can cause an excessive intracranial overheating. Repeating such behaviour on a daily basis, it is no wonder how tumours and carcinomas can be developed even by a healthy organism.

The most important sources of EMF pollution in a household environment are:

- WiFi and cordless: these communication bases emit quite a strong radiation (1-5 V/m) approximately at 1m or 3.2ft away from it;
- A microwave oven: the faraday cage does not protect from radiations when in usage, so that the local EMF is in the range of (10-20 V/m);
- New smart meters for electric, water and methane: they constantly communicate with the distributors to give usage data.

Possible countermeasures include EMF blocking paint and Orgone purifiers as the back effect of such invisible pollution is the creation of dangerous level of Deadly Orgone – DOR.

AIR POLLUTION

For hundreds of millions of people in urban areas around the globe, the air they inhale is so saturated with contamination that it is actually shortening their lives. Occupants of Los Angeles, for instance, lose almost a year of life because of dirty air. In Venice, Italy, this number raises to 1.7 years. In any case, by a wide edge, inhabitants of urban areas all through Asia pay the highest cost. If present air-contamination continues, the normal individual in significant Asian urban areas like Beijing, Lahore and Delhi will live more than five years less than if their air met guidelines set up by the World Health Organization.

Those are the discoveries of the Air Quality Life Index (AQLI), another tool delivered by the Energy Policy Institute at the University of Chicago. The AQLI combines frontier look research in wellbeing and financial aspects with about two decades of satellite-derived estimations of particulate matter contamination for the whole world. Therefore, the Index permits individuals anyplace on Earth to get to data about the air they breathe and how it affects their health in the metric that matters most: its impact on their future life expectancy.

The acquired data is striking: the AQLI distinguishes air pollution as the single most greatest danger to human wellbeing globally, with its toll on life expectancy exceeding that of contaminated ailments and uncommon diseases, for example, malaria and tuberculosis and even behavioural dangers like cigarettes, given current smoking rates especially among youngsters.

In spite of the fact that some particulate matter is produced by common sources, for example, dust, the primary source of urban particulate matter contamination is energy use. The petroleum derivatives that power 80% of the global energy framework, including transportation, industry and electric power, produce particulate issue contamination during combustion, releasing it into the atmosphere. From that point, the small particles enter our lungs, where they can circumvent the body's natural defences and spread well past our respiratory system.

CHEMICAL POLLUTION

Waste from factories is unfortunately often directly disposed of in water, causing a massive pollution of the whole environment, starting from the flora to the fauna. Water has got the peculiar ability to pass through almost all materials and rocks, so that dissolved pollutants easily spread over very vast areas.

Artificial contamination of seas and rivers can influence human wellbeing by direct poisoning of food and water sources.

The most dangerous potential threats and huge scale consequences for human health is related to the polluted fish. Toxins like dioxins and PCBs, mercury and heavy metals, the typical pathway to the body is through eating, from fish specifically (Swedish Environmental Protection Agency, 2009). Ecological contaminant has been connected to an increased number of cancer and allergies, hypersensitivity, early neurological degradation, and a troublesome influence of the thyroid and the endocrine system. It is not just water to be dumped with chemical contaminant agents, but also air.

An aspect that requires further revision is the transformation of dissolved minerals into highly charged colloidal particles. It seems to imply that, although chemicals are not actually eradicated, they might be transformed into a state in which they are less harmful to the human being. The consumption of Orgone-enriched water strengthens the immune system and should make the body handle much better the detrimental effects of external pollutants, chemicals and electromagnetic radiation.

MENTAL POLLUTION

With the sudden introduction of new technologies that constantly try to grab people's attention with notifications, light and sounds, individuals are subject to an effect called *mental contamination* (or *mental pollution*). Ads, omnipresent advertisement, constant repetition and mental tricks used in all sorts of technologies, from social networks to dating apps, human mind is clueless and easily trapped by perfectly designed systems able to grab attention and generate higher revenue. Like the way water, air, and soil contamination compromises human heath, mental pollution may be the reason for poor memory and a difficulty in learning. Younger generations are very affected being born in this constant stream of shiny lights, beeps, and voices creating what has been defined as a "distracted mind".

A PHILOSOPHICAL BRIEF

Why is talking about life force so important? Well, it's a long story starting from antiquity and from the concepts of chi and chakra, going through Wilhelm Reich's Orgone to arrive at the theory of everything of the physicist Nassim Haramein: the fil rouge is the concept of universal energy, which permeates the entire cosmos. Self-actualization is an infinite life-long journey on a dangerous path made of futile joy and sterile emotions. Internalisation of emotions is the key to stability. At the beginning of human history, people have been intermingling with the nexus of Universe via homomorphic resonance, i.e. connecting with entities of the same form (organic beings with organic beings, animals with animals and plants).

The mindset stays in an ever-present ennobling of inspiration, inner-struggle and empathy that may amplify our connection to the planet itself or detach our physical body from the stream of energy all around. The key is psychological stability of the subject, as well as the ability to maintain a neutral emotional status towards positive and negative events, welcoming the first and accepting the latter.

About it Reich wrote that: "The emotions [...] are

the experiential material of mysticism. Therefore, the narrow-minded mechanist concludes that those who deal with emotions are mystics. The understanding of emotions is so remote to the mechanist's thinking that there is no room for it in his natural scientific investigation".

And later on, in *Character and Society*: "The most general result for character formation is the mutual lack of contact and the replacement of natural human relationships by artificial, formalistic relationships. In spite of our collective living, there are only rare individuals who are not fundamentally lonely, empty, and superficial. This psychic situation creates the longing for *release* or even for *dissolution*. People have lost the capacity for freely swimming in the current of life".

Science cannot deeply investigate the real essence of emotions as they are intangible by nature, a stream of thoughts, ideas, images and passion within the individual that cannot be reduced to a single biochemical stimulus. In fact, if that were the case, drugs could be developed to perfectly reproduce the whole plethora of emotions of humans. Yet, as shown by the devastating effects of illicit drugs used as stimulants, they can only mock a tiny part of a real emotion, creating strong limited sensations of pleasure or pain relief in a way that destroys the organisms. Thus, any form of resistance, self-incrimination and despair will result in an internal imbalance with detrimental health effects.

REALISM VS IDEALISM

If one examines cultural appropriation of sex as a dirty and impure act, one is faced with a choice: either reject the sub constructivist paradigm of discourse or conclude that erotic love still has a relevant significance. Lacan uses the term "post semiotic Marxism" to denote a postcapitalist deconstructivist theory stating an intrinsic meaning to life. Likewise, Sontag suggests the use of social realism to attack the status quo.

In his works, a predominant concept is the idea of pre-cultural culture, the type of knowledge existing before the invention of modern society. The subject is deeply interpolated into a semantic logic that foresees sexuality as a fundamental part of life. Thus, an abundance of deconstructions concerning social realism may be unrevealed. Such premise implies that context is a product of the masses, but only if truth is interchangeable with narrativity; if that is not the case, we can assume that the importance of the individual is his/her significant form rather than his/her subjectivity.

Looking at Diderot reading of his contemporary society, he concluded that government is totally capable of social modification, also considering Foucault's analysis of neo-capitalist rationalism: rationale is a product of the government, justification is the one of the masses. Any form of conceptualism concerning the bridge between society and sexual identity is doomed to be marked as illogical. The erotic subject is contextualised into an experience that revolves around the idea of language as a standalone reality, yet erotic language is highly censored and misunderstood often leading to exacerbation or exaggeration falling into pornography or the pure hedonism. It is very evident how the common ground between sexuality and social class was well-established into the Christian culture whereas ancient Greeks and Romans saw sex in a completely different way.

STEPS TOWARDS AN ORGONOMIC SOCIETY

Given the connection between the individual and society, improvements in a person's life and wellbeing include changes of their depiction to the public eye. The way a person is treated can be very useful for the individual status, however it is insignificant for the general public. Social, not politically ordered, programs are expected to invert the harsh parts of present-day culture and introduce progressive changes that endorse a dominant power changing too rapidly, bring chaos and a considerably more prominent increment in insensitive oligarchic supremacies. Thus, society by and large can't work in an unarmored style at present (in spite of the fact that the single individual can), and absolutely not until the sex-shaming, life-limiting parts of culture change. Opportunity merchants are dangerous. Change must come step by step.

To build up a social orgonomy program and empower orgonomic study of issues in the social sector is to come back to Reich's initial work in Europe.

There, in research hubs, he effectively followed and helped several individuals in regards to social, psychoaffective and familiar issues in their day by day lives. On one level, making a viable social program, one that really helps people, includes whole-hearted contact with the patient, intellectual connection, and proactive, yet prudent, intercession. These qualities are hard to find. There are, however, programs in the nation and around the globe that are reacting in a positive way to these social issues: a shelter for maltreated women; a gathering home for orphans and teenagers; some preschool programs; local microcredit programs; maternity care, and domiciliary medical support.

Nevertheless, social support programs are often set up with next to zero logical understanding of what they are doing, without real comprehension of the human condition and, as expected, fail to precisely evaluate issues and to change current conditions and straightforwardness human anguish.

Social orgonomy tries to understand the human condition, the fervent plague, life-limiting and sex-shaming culture, and offers a hypothetical viewpoint to distinguish the underlying reasons of and answers for social issues. *Orgonometry,* a utilitarian perspective, permits investigation of complex issues in different parts of social living.

Changing society to promote a healthy mindset, touchy care of newborn children and freedom of sexual expression were Reich's main social concern. Beginning with pre-birth and pregnant women living conditions, care of newborn children is the trickiest and most fundamental part of society that needs to change so as to improve human misery and social despair. Supporting the supremacy and the closeness of mother-infant contact, perceiving the affectability and flawless enthusiasm of newborn children, and putting no useless adjustments in front of the infant care are inalienable basic rights.

Social orgonomy can be used to create the hypothesis of social intercession in addition to study and build the working adequacy of the social program once properly planned. Absence of compelling execution of social projects adds to disappointment around the world. Working connections inside an association, and correspondence and division of work, fundamental to any working gathering, might be contemplated from a social orgonomic point of view.

Expand complexities, laws, administrative work, and principles by which *great* is to be cultivated frequently slaughter programs.

An educator as of late remarked that the nature of present-day life is decreased by our unhinged, frenzied life style and that accommodating programs that are enacted and allocated to instructors just add to the quick, hampered pace of life and leave less time for the instructor, tyke, parent, and network to converse with one another in a significant way. Such projects hinder commitment and occupy consideration from issues and arrangements. Social projects can expand society's working in the zones of adoration, work, and information the three zones Reich called "the wellsprings of life". It is additionally accurately these capacities that are the focal point of enthusiastic plague assaults, and any social program should address the enthusiastic plague. The passionate plague comes in varying structures: out and out defamation, individual and property annihilation, oppression; and increasingly unobtrusive however horrible allusion, tattle, legitimate declarations, formality, implicit deterrent, well-voiced issues of political rightness, and bureaucratic principles. The previously mentioned program in South Carolina initially made condoms accessible to adolescents at school yet government standards ceased that. Undaunted what's more, pushing ahead, the chief offered condoms to the nearby barbershop where the hair stylist keeps them on a first rate and offers them to whomever makes a trip and asks, and utilizes the chance to draw in the adolescents in discussion about their insight into condoms and their objectives and expectations throughout everyday life.

VITANISM

Reich's Orgone energy is an idea inspired by vitalism, and becomes central to his work in the 1930s and 1940s. While it has significant parallels to earlier ideas of vital energy, orgone germinates as a variant of Freud's theory of the libido. One author sees orgone as a form of *libido unbound* and notes that Reich essentially "conceptualises energy as entirely sexual, and pursues a quantitative approach to the libido".

In *Civilization and its Discontents* (1930) Freud argues that healthy biological urges (the libido) are suppressed or sublimated to the demands of the social order (with bourgeois morality at its core). Seeing the libido as a kind of *"life force"* Reich responded critically to Freud's conception.

In his hundred-page monograph, Freud presented his pessimistic evaluation of the eternal conflicts among the libido, ego, and superego within the self and between the self and its community. *Eros and death* remained the same major players they were nine years earlier (in *Beyond the Pleasure Principle*), and the aggressive drive assumed center stage. Needless to say, none of this sat well with the antidualist Reich, who thought that there was no death drive and that the so-called drive for aggression was the result of bodily armouring rather than an innate piece of nature.

In the early 1920s the two thinkers were in closer concert, but quickly diverged as Reich developed a pathological vision of modern society: "Indeed, at its most extreme, orgonomy turned against the Freudian virtues of sublimation, strength of character, and self-knowledge, abominating them as toxic substances, literally carcinogens". Reich explored repression that marked people in deep, physically challenging ways to overcome the self-essence of individuals abstracted from the norming processes of society itself encapsulating individuals within a cage to uniform everything according to pre-defined abstract principles. More on this topic in the book about Vitanism by the same author.

GLOSSARY

5G: a new technology enabling faster communications and transmission data rate but with higher energetic emissions and possible negative effects for human health

Allopathic medicine: a terrible drift in modern medicine aimed at treating symptoms ignoring the original triggering causes

Bion: a new subatomic particle highly interactive with organic compounds

Chi: an universal force regulating bodily functions giving "life" to creatures according to traditional Chinese medicine and philosophy

Diathermy: a device that heats tissues using high frequencies electric currents

DOR: Deadly or Negative Orgone, a form of energy showing detrimental effects created by the degradation of Positive Orgone

EMF: electromagnetic frequency, usually referred as the spectrum of emissions of a certain device

EMP: electromagnetic pollution caused by multiple emitting sources and probably affecting human health

Energy: a quantitative property to be transferred to an object in order to perform some kind of work, whether kinetic, thermal or internal

Ether: a medium filling empty space allowing electromagnetic waves (thus light) to propagate

Kirlian photography: a collection of photographic techniques used to capture the phenomenon of electrical coronal discharges able to depict Orgone energy

Magnetostriction: the property of ferromagnetic materials that causes them to change their shape or dimensions during the process of magnetisation

ORAC: the original Orgone Accumulator by Reich made of several layers of organic and inorganic materials

ORANUR: the adverse reaction of Orgone when put in presence of radioactivity

Orgone: a concept developed by Wilhelm Reich to explain a new type of life energy

Orgonite: a device able to capture and purify

environmental Orgone as well as reduce electromagnetic pollution

Phosphorescence: a physical phenomenon of spontaneous emission of light on a longer timescale than fluorescence

Piezoelectricity: the electric charge built up in certain solid materials when mechanical stress is applied

Polarisation: the ability of waves to oscillate in more than one direction or plane

POR: Positive Orgone, with beneficial effects on the environment and living organisms

Prana: a life-giving universal energy which flows in currents inside and outside the body according to Hinduism

Quantum physics: a physical theory to describe matter in terms of quantised packets of energy exchanged by atoms and subatomic particles, built around the fundamental dogma of uncertainty and probabilistic results over any direct measure

Scintillators: materials that luminate when excited by ionising radiation/particles, often used in physics to detect incoming particles

WIMP: weakly interactive massive particle, hypothetical particles that are one of the proposed candidates for dark matter

BIBLIOGRAPHY

Theory of microlepton fields, Anatoly F. Ohatrin

https://medium.com/@leopoldo.orsini/the-orgonite-scam-8bd0fb059429, *The Orgonite Scam* by Leopoldo Orsini Corvetti

http://www.orgonelab.org, *Orgone Biophysical Research Lab* (OBRL), James DeMeo

Vitanìa, l'azione vitalizzante endogena (Vitanism) by Leopoldo Orsini Corvetti

The Function of the Orgasm, Reich 1942 (Die Entdeckung des Orgons Erster Teil: Die Funktion des Orgasmus, translated by Theodore P. Wolfe)

The Bioelectrical Investigation of Sexuality and Anxiety, Reich 1937

Character Analysis, Reich 1945 (Charakteranalyse, translated by Theodore P. Wolfe)

The Sexual Revolution, Reich 1945 (Die Sexualität im Kulturkampf, translated by Theodore P. Wolfe)

The Mass Psychology of Fascism, 1946 (Massenpsychologie des Faschismus, translated by Theodore P. Wolfe)

The Discovery of Orgone & The Cancer Biopathy, Reich 1948

Listen, Little Man!, 1948 (Rede an den kleinen Mann, translated by Theodore P. Wolfe)

The Orgone Energy Accumulator, Its Scientific and Medical Use, Reich 1948

The Orgone Accumulator Handbook: Wilhelm Reich's Life-Energy Discoveries and Healing Tools for the 21st Century, with Construction Plans, James DeMeo 2010

Ether, God and Devil, Reich 1949

Cosmic Superimposition: Man's Orgonotic Roots in Nature, Reich 1951

The Invasion of Compulsory Sex-Morality, Reich 1951

The Oranur Experiment: First Report (1947–1951), Reich 1951

People in Trouble (The Emotional Plague of Mankind), Reich 1953 (Menschen im Staat)

The Einstein Affair, Reich 1953

Contact with Space: Oranur Second Report (1951–1956), Reich 1957

The Bion Experiments: On the Origin of Life, Farrar, Straus and Giroux 1979 (Die Bione: Zur Entstehung des vegetativen Lebens)

Genitality in the Theory and Therapy of Neurosis, Farrar, Straus and Giroux, 1980

The Bioelectrical Investigation of Sexuality and Anxiety, Reich 1982

Children of the Future: On the Prevention of Sexual Pathology, Reich 1983 (the chapter entitled "The Sexual Rights of Youth" is a revision of Der Sexuelle Kampf der Jugend)

The Distracted Mind: Ancient Brains in a High-Tech World, Adam Gazzaley and Larry D. Rosen

HEILEN MIT ORGONENERGIE: Die Medizinische Orgonomie, Kavouras 1997 (translated HEALING BY ORGONE ENERGY: Medical Orgonomy)

Wilhelm Reich: the man who invented free love, Turner, Christopher 2011

The Body in Psychotherapy, Edward W. L. Smith 2000

Me and the Orgone. American College of Orgonomy, Bean 2000

The Historic Context of Reich's Laboratory Work, James Strick

NRHP nomination for Orgone Energy Observatory, National Park Service

Freud's Free Clinics: Psychoanalysis & Social Justice, Danto, Elizabeth Ann 2007

"Wilhelm Reich", Architects of the Culture of Death, DeMarco, Donald and Wiker, Benjamin D. 2004

Heretic's Notebook: Emotions, Protocells, Ether-Drift and Cosmic Life-Energy, with New Research Supporting Wilhelm Reich, James DeMeo 2011

In Defense of Wilhelm Reich: Opposing the 80-Years' War of Mainstream Defamatory Slander Against One of the 20th Century's Most Brilliant Physicians and Natural Scientists, James DeMeo 2012

The Greatness of Wilhelm Reich, Edwards, Paul 1977

Bibliography on Orgone Biophysics: Selected Citations of Books and Articles by Wilhelm Reich and Others on the Orgone Energy, James DeMeo 1986

INDEX

Printed by Lulu Press inc. — First edition December 2019
Graphics and images are either Open Source or Creative Commons ® license
ISBN 978-0-244-54393-8

Printed in the USA